FA–220

Firefighter Fatality
Retrospective Study

April 2002

Prepared by

**TriData Corporation
1000 Wilson Boulevard
Arlington, VA 22209**

for

**Federal Emergency Management Agency
United States Fire Administration
National Fire Data Center**

This publication was produced under Order No. EME-2000-DO-0396 by TriData Corporation for the Federal Emergency Management Agency, United States Fire Administration

UNITED STATES FIRE ADMINISTRATION
MISSION STATEMENT

As an entity of the Federal Emergency Management Agency (FEMA), the mission of the United States Fire Administration (USFA) is to reduce life and economic losses due to fire and related emergencies, through leadership, advocacy, coordination, and support. We serve the nation independently, in coordination with other federal agencies, and in partnership with fire protection and emergency service communities. With a commitment to excellence, we provide public education, training, technology, and data initiatives.

CONTENTS

List of Figures v

List of Tables vi

Preface vii

Acknowledgments viii

Executive Summary 1

    Introduction and Objectives 1

    Findings 1

    Prevention and Resources 3

Introduction 5

    Goals and Objectives 6

Methodology 9

    Data Sources 9

    Development of Database 10

    Inclusion Criteria 10

Findings 13

    Trend in Firefighter Fatalities 13

    Type of Incident 13

    Affiliation of Firefighter Fatalities and Type of Agency 14

    Rank of Firefighter 16

    Gender 18

    Age 18

    Type of Duty 20

    Motor Vehicle Collisions 21

    Immediate Cause and Nature of Fatal Injury 23

    Comparison to Other Fatality Rates 25

    Time of Injury 26

    Fixed Property Use 27

    Cause of Fire 28

    Geographical Distribution of Firefighter Fatalities 29

    Comparison of Florida With Pennsylvania 36

    Multiple Firefighter Fatality Incidents 37

    Self-Contained Breathing Apparatus Depletion 39

    Personal Alert Safety System Device Activation 39

    Wildland Firefighters 39

Prevention of Firefighter Fatalities 41

    Pre-Existing Conditions 41

    Training 45

    Fireground 46

    Motor Vehicle Collisions 48

Resources **51**

    Fire Service Resources 51

    Financial Assistance for Children and Spouses 54

Appendix—Documentation **57**

    Entity Relational Diagram 57

    Field Names and Description of Data 58

    Data Dictionary 62

    Rules for Coding Data 73

References **75**

LIST OF FIGURES

1   Trend in Firefighter Fatalities (1977–2000)   5

2   Firefighter Fatalities (1977–2000)   13

3   Types of Incident Resulting in Fatalities (1990–2000)   14

4   Fire-Incident-Related Firefighter Fatalities
    per 100,000 Incidents (1977–2000)   14

5   Firefighter Fatalities by Affiliation (1990–2000)   15

6   Career–Volunteer Distribution (1990–2000)   15

7   Type of Agency with Firefighter Fatality (1990–2000)   16

8   Firefighter Fatalities by Rank (1990–2000)   17

9   Rank of Firefighter vs. Type of Duty at Time of Fatal Injury
    (1990–2000)   17

10 Gender of Firefighter Fatalities (1990–2000)   18

11 Age at Time of Injury (1990–2000)   18

12. Type of Duty by Age of Firefighter (1990–2000)   19

13 Type of Duty at Time of Fatal Injury (1990–2000)   20

14 Emergency vs. Non-Emergency Duty (1990–2000)   21

15 Leading Types of Training Activities Associated With Fatalities
    (1990–2000)   22

16 Type of Vehicle Involved in Collision (1990–2000)   22

17 Decedent's Location in Vehicle (1990–2000)   23

18 Immediate Cause of Fatal Injury (1990–2000)   23

19 Nature of Fatal Injury (1990–2000)   24

20 Percent of Heart Attack Deaths by Year (1984–2000)   24

21 Comparison of Type of Duty for Heart Attack With Non-Heart Attack
    Fatalities (1990–2000)   25

22 Percent of Injuries by Time of Day (1990–2000)   26

23 Percent of Traumatic vs. Heart Attack Deaths by Hour of the Day
    (1990–2000)   27

24 Fixed Property Use of Incident Where Injury Occurred (1990–2000) . .   28

25 Cause of Fire, If a Fire Incident (1990–2000)   29

26 Map of Firefighter Fatalities, by State of Affiliation (1990–2000)   31

27 Firefighter Fatalities, by Affiliation Per 10 Million Population   32

28 Firefighter Fatalities, by Type of Incident Per 10 Million Population . . .   33

29 Entity Relational Diagram, Firefighter Fatality Database   57

LIST OF TABLES

1 Leading Nature of Fatal Injury by Age (1990–2000) 19

2 Emergency vs. Non-Emergency Duties by Age of Firefighter (1990–2000) 20

3 Comparison of Heart Attack Fatalities by Occupation (1990–2000) ... 26

4 Time of Injury (1990–2000) 27

5 Firefighter Fatalities by State and Per 10 Million Population (1990–2000) 30

6 Type of Incident by State (1990–2000) 35

7 Affiliation by State (1990–2000) 35

8 Nature of Fatal Injury by State (1990–2000) 35

9 Firefighter Fatality Data Elements: Florida and Pennsylvania 36

10 Multiple Firefighter Fatalities 38

11 Nature of Fatal Injury, Single- vs. Multiple-Fatality Incidents 38

12 Nature of Fatal Injury, Wildland Firefighters vs. Non-Wildland Firefighters (1990–2000) 40

PREFACE

On September 11, 2001, terrorists attacked New York City's World Trade Center and the Pentagon in Arlington, Virginia. Following the initial attack, both towers of the World Trade Center collapsed, killing thousands of civilians, dozens of police officers, and hundreds of firefighters. Among fatalities from the Fire Department of the City of New York (FDNY) were the Chief of the Department, First Deputy Commissioner, Chief of Special Operations, and one of the Department's Chaplains.

The World Trade Center disaster represents the largest loss of firefighters in a single incident in the United States since 1947, when 27 firefighters perished in fires and explosions aboard two Texas City ships.

According to Harold Schaitberger, president of the International Association of Fire Fighters, "This was the darkest day in the history of the firefighters of the world. It will change every one of our lives forever."

The men and women of America's fire service who responded to these attacks have brought honor upon themselves and the entire fire service community through their heroic actions and commitment to duty. For those that made the supreme sacrifice in the course of their actions, may their souls rest in peace.

This report is dedicated to the families of all firefighters who have died while on duty. Through the lessons learned from their passing, it is hoped that future lives will be saved.

ACKNOWLEDGMENTS

This study of firefighter fatalities would not have been possible without the cooperation and assistance of many members of the fire service throughout the United States. Additionally, the staffs of the National Fallen Firefighters' Foundation (NFFF), Public Safety Officer's Benefit (PSOB) program, and National Fire Protection Association (NFPA) provided data and other information that were invaluable in the preparation of this report. We thank them for their contributions to this project.

TriData Corporation conducted this analysis under contract EME–2000–DO–0396 for FEMA, USFA.

EXECUTIVE SUMMARY

Each year in the United States and its protectorates, approximately 100 firefighters are killed while on duty and tens of thousands are injured. Although the number of firefighter fatalities has steadily decreased over the past 20 years, the incidence of firefighter fatalities per 100,000 incidents has actually risen over the last 5 years, with 1999 having the highest rate of firefighter fatalities per 100,000 incidents since 1978.

Introduction and Objectives

In the last decade, several high-profile incidents involving firefighter fatalities have brought national attention to the issue of firefighter mortality in the United States. While the attention from the national media has been fleeting, the awareness of the continued high level of fatalities has changed the fabric of the fire service and prompted many organizations and fire departments to initiate programs to protect firefighters.

This analysis sought to identify trends in mortality and examine relationships among data elements. To this end, data were collected on firefighter fatalities between 1990 and 2000. (For further information, see the "Methodology" section or the Appendix.) Using this analysis, better targeted prevention strategies can be developed in keeping with the USFA's goal to reduce firefighter deaths 25 percent by 2005. In contrast to the annual USFA firefighter fatality reports, this analysis allowed for comparisons over time to determine any changes in firefighter mortality, with a depth of scrutiny not present in earlier analyses.

Ultimately, some forces and circumstances that lead to firefighter fatalities are simply beyond human control. However, through research, study, training, improved operations, development of new technologies, the appropriate use of staffing, and other factors, it should be possible to significantly reduce the number of firefighters killed each year.

Findings

Nature of Fatal Injury—The leading nature of fatal injuries to firefighters is heart attack (44 percent); trauma, including internal and head injuries, is the second leading type of fatal injury at 27 percent. Asphyxia and burns combined account for 20 percent of fatalities. More firefighters die from trauma than from asphyxiation and burns combined.

Firefighters under the age of 35 are more likely to be killed by traumatic injuries[1] than they are to die of medical causes (e.g., heart attack, stroke). After age 35, the proportion of deaths due to traumatic injuries decreases, and the proportion of deaths due to medical causes rises steadily.

---

[1]Traumatic injury means a wound or the condition of the body caused by external force, including injuries inflicted by bullets, explosives, sharp instruments, blunt objects or other physical blows, chemicals, electricity, climatic conditions, infectious diseases, radiation, and bacteria, but excluding stress and strain [Ref. 1].

Age—Approximately 60 percent of firefighter fatalities were over the age of 40 when they were killed, and one-third were over 50. Nationally, firefighters over the age of 40 comprise 46 percent of the fire service, with those over 50 accounting for only 16 percent of firefighters. Although older firefighters possess a wealth of invaluable knowledge and experience, they are killed while on duty at a rate disproportionate to their representation in the fire service. Also, these older firefighters tend to be affiliated with volunteer agencies. About 40 percent of volunteer firefighters are over the age of 50, compared to only 25 percent of career firefighters.

Affiliation—The majority of firefighter fatalities (57 percent) were members of local or municipal volunteer fire agencies (including combination departments, which are comprised of both career and volunteer personnel). Full-time career personnel account for 33 percent of firefighter fatalities; they comprise only approximately 26 percent of the American fire service. Numerically more volunteer firefighters are killed than career personnel, yet career personnel are killed at a rate disproportionate to their representation in the fire service.

Emergency Medical Services (EMS) Fatalities—In many fire departments, EMS calls account for between 50 and 80 percent of emergency call volume. These incidents result in only 3 percent of firefighter fatalities. Trauma (internal/head) accounts for the deaths of 50 percent of firefighters who were involved in EMS operations at the time of their fatal injury; another 38 percent involved in EMS operations died from heart attacks.

Type of Duty—Of those firefighters killed while en route to an incident, 85 percent were volunteers. For firefighters killed performing in-station duties, 69 percent were career personnel; the majority of those deaths were the result of heart attacks. These variations can be attributed to differences between career and volunteer agencies. Generally, unless they are on a call or other fire department business, career personnel are required to be in the fire station for the duration of their shift, which is generally between 10 and 24 hours long. As a result, volunteers are more likely than career firefighters to die while responding.

Motor Vehicle Collisions (MVCs)—Since 1984, MVCs have accounted for between 20 and 25 percent of firefighter fatalities annually. One quarter of firefighters who died in MVCs were killed in private/personally owned vehicles (POVs). Following POVs, the apparatus most often involved in fatal collisions were tankers, engines/pumpers, and airplanes. More firefighters are killed in tanker collisions than in engines and ladders combined.

About 27 percent of fatalities killed in MVCs were ejected from the vehicle at the time of the collision; only 21 percent of firefighters were reportedly wearing their seatbelts prior to the collision.

Most volunteer departments do not require personnel to stand by in the fire station; members are allowed to respond directly to incidents from their homes or workplaces, often in their POVs. As a result, volunteers are more likely than career firefighters to be killed in POV collisions. Moreover, they are more likely to be involved in collisions involving tankers, which are predominantly used in rural areas without hydrants or other readily available sources of water. Such areas are almost exclusively protected by volunteer fire departments.

Training—In the last decade, approximately 6 percent of firefighter fatalities occurred during training activities, a larger proportion than in the previous decade. Over time, the leading type of training activity resulting in fatalities has remained physical fitness, followed by equipment/apparatus drills and live fire exercises.

Multiple Firefighter Fatality Incidents—Between 1990 and 2000, 8 percent of fatal incidents involved the death of more than one firefighter; these incidents accounted for 18 percent of firefighter fatalities. About 14 percent of firefighters were killed in incidents that resulted in the deaths of two or three firefighters. Incidents involving the death of more than four firefighters are rare, and accounted for only 3 percent of fatalities. These findings represent an increase from an earlier USFA study that found that between 1982 and 1991, only 4 percent of incidents involved the death of more than one firefighter; those incidents accounted for 13 percent of firefighter fatalities.

Approximately 90 percent of firefighters killed in multiple-fatality incidents die of traumatic injuries. In contrast, only 37 percent of those killed in single-fatality incidents die from traumatic injuries.

Prevention and Resources

Some circumstances that lead to the deaths of firefighters are simply beyond human control. Generally, however, most firefighter fatalities are the result of a chain of events, which, if detected early, has the potential be broken and prevent many, or even most, fatalities.

Prevention strategies discussed include increased emphasis on physical fitness, dietary changes, behavior modification, changes in operational strategies and tactics, and more stringent adherence to standard operating procedures (SOPs).

Resources are available from a variety of federal, state, local, and private agencies. Information and contacts are included for specific programs designed for fire departments that experience a firefighter fatality and the firefighter's surviving family.

INTRODUCTION

The deaths of firefighters profoundly affect not only the families they leave behind, but also the communities in which they lived, the firefighters with whom they served, and the fire service as a whole. Each year in the United States and its protectorates, approximately 100 firefighters are killed while on duty and tens of thousands are injured. As depicted in Figure 1, the incidence of firefighter fatalities has trended downward (38 percent) over the past 25 years, from a high of 171 in 1978 to a low of 77 in 1992.

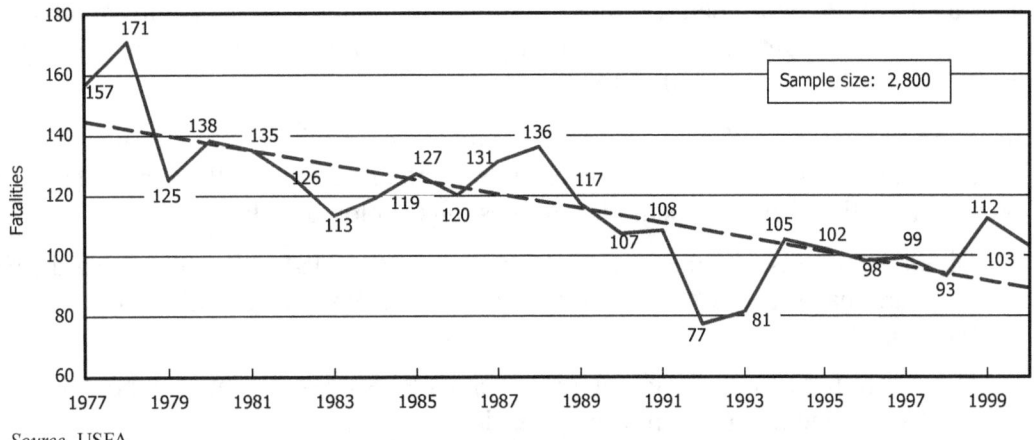

Source USFA.

Figure 1. Trend in Firefighter Fatalities (1977–2000)

The number of firefighter fatalities annually differs slightly depending on the criteria used to define an on-duty fatality. It is possible that a firefighter fatality could be declared eligible as an on-duty death some years after the firefighter's injury. As a result, it is not uncommon to find fluctuations of from one to five fatalities annually, depending on the methodology used to compile the report.

In the last decade, several high-profile incidents involving firefighter fatalities have brought national attention to the issue of firefighter mortality in the United States (e.g., six firefighters killed in Worcester, Massachusetts, in 1999 and 14 killed at Storm King Mountain, Colorado, in 1994). Also, the events of 9/11 have dramatically highlighted the heroic activities of the United States' fire service community. A growing awareness of the continued level of fatalities has changed the fabric of the fire service and prompted many fire departments and fire service organizations to initiate programs to protect firefighters. Fire departments throughout the country are adjusting their tactics to promote firefighter safety and reduce firefighter deaths and injuries. One such adjustment is the designation of rapid intervention teams/crews (RITs/RICs) for working fires. (These teams or crews stand by outside of a structure and are deployed immediately to initiate a rescue attempt after a firefighter calls for help or is declared missing.)

Also to reduce firefighter mortality, organizations such as the International Association of Fire Fighters (IAFF), International Association of Fire Chiefs (IAFC), National Volunteer

Fire Council (NVFC), and the United States Fire Administration (USFA) have begun to develop additional programs or new initiatives intended to promote firefighter health, safety, and well being. Similarly, efforts are underway to develop training programs to teach firefighters how to rescue themselves and their fellow firefighters in the event they become trapped or disoriented in a fire.

Ultimately, through research, study, training, improved operations, development of new technologies, the appropriate use of staffing, and other factors, it should be possible to substantially reduce the number of firefighters killed each year.

Goals and Objectives

This analysis sought to identify trends in mortality, examine relationships among data elements, and aid in targeting prevention strategies for the USFA's goal to reduce firefighter deaths 25 percent by the year 2005 [Ref. 2].

In contrast to the annual USFA firefighter fatality reports, this analysis allowed for comparisons over time to determine any changes in firefighter mortality, with a depth of scrutiny not present in earlier analyses. Examples of questions explored included:

- Given the increase in emergency medical services (EMS) call volume over the past 20 years, has the number of firefighter fatalities associated with EMS calls also increased?

- Has the introduction of technologies such as PASS devices and integrated PASS/SCBA affected the trends in firefighter deaths?

Examples of relationships in the data explored in this analysis include changes to death rates (or the magnitude of deaths) due to enhancements such as Self-Contained Breathing Apparatus (SCBA) or Personal Alert Safety System (PASS). The analysis also considered patterns in the deaths of career vs. volunteer firefighters and the relationship among age and gender and the cause of firefighter deaths.

The report includes a "Resources" section for fire department reference in the event of a firefighter fatality and for firefighters' surviving spouses and children. These benefits are available through federal, state, local, and private sources, including labor organizations. The goal is to provide the fire service with resources for developing a plan to deal with on-duty fatalities as well as an understanding of what is available and where to turn for help following the death of a firefighter.

Future Analyses—Analysis of the current database provides a broad perspective on historical trends in firefighter fatalities; however, there are some areas where the data are not currently available to perform quality analyses. For example, in 1999 the Occupational Safety and Health Administration (OSHA) revised its standard on respiratory protection (29 CFR 1910–134), known as Two-In/Two-Out. It is not yet clear how the policy's modification will affect firefighter deaths. Similarly, training program improvements, the development of health and wellness initiatives, and the use of RITs may also affect future trends in firefighter injuries and deaths.

As the database is expanded and updated, it should become possible to more clearly determine the effects of these changes and trends in the fire service. These issues will be addressed by future USFA publications.

METHODOLOGY

Data Sources

The report is based on data from a number of sources. The primary sources are USFA files from the National Fire Data Center (NFDC), the National Fallen Firefighters' Foundation (NFFF), and the Public Safety Officer's Benefit (PSOB) program.

The NFDC describes the nation's fire problem, proposes possible solutions and national priorities, monitors resulting programs, and provides information to the public and fire organizations. The NFDC files contain detailed lists of firefighter fatalities, individual files on some fatalities, and annual reports with summary data for firefighter fatalities in a given year. Although the NFDC files are extensive, they are not complete; not all firefighter fatalities are enumerated and files contain varying levels of information. In particular, information on firefighter fatalities in the early part of the 1990s is sparse; the level of detail increases toward the end of the decade.

The NFFF collects data on firefighter fatalities to determine if the fatality meets the criteria for inclusion on the National Fallen Firefighter's Memorial at the National Emergency Training Center (NETC) in Emmitsburg, Maryland. Eligible firefighters include those who meet PSOB guidelines (whether or not the firefighter's next of kin complete the application process) or deaths from injuries, heart attacks, or illnesses directly attributable to a specific emergency incident or training activity. Private firefighters such as those in industrial brigades are included, provided the deaths meet certain standards. Some types of fatalities are excluded (e.g., deaths attributable to suicide, alcohol, or substance abuse). These excluded cases account for a very small fraction of firefighter fatalities.[2]

Files maintained by the PSOB program provided another significant source of information on firefighter fatalities. The PSOB program is administered by the Department of Justice and provides a monetary benefit to the survivors of a public safety officer who dies while on duty. Public safety officers are defined as career (full time) or volunteer (part time) firefighters, law enforcement officers, and emergency medical workers. To qualify for benefits, the circumstances surrounding the death must be traumatic in nature (e.g., smoke inhalation, structural collapse, motor vehicle collision, gunshot wound). Deaths that are nontraumatic in nature (e.g., heart attacks, strokes) do not generally qualify for the PSOB benefit. To process a claim, PSOB requires the officer's surviving family and employer to submit certified documents supporting their application. Required documents that were collected for this report include a detailed description of the circumstances surrounding the officer's death, a copy of the autopsy and toxicology reports, and a copy of the agency's incident report. These files are quite comprehensive for traumatic firefighter fatalities; however, they contain extremely limited information regarding nontraumatic fatalities.

---

[2]These eligibility requirements were determined at a meeting on June 18, 1997, which included representatives from the major fire service organizations who unanimously agreed to adopt the new criteria as listed here retroactive to January 1, 1997.

Summaries of each fatal incident from 1994–2000 were considered in the analysis. However, given their collective length, they were not included in the text of the report. Rather, they will be made available on the Internet by the USFA at http://www.usfa.fema.gov.

Other sources of data included trade journals, news reports, fatality investigations by the National Institute for Occupational Safety and Health (NIOSH), journal articles, and Internet sites.

Development of Database

A key product of the research process was the development of a multiyear, relational firefighter fatality database. The database comprises seven tables. For further information, see the Appendix.

Inclusion Criteria

Firefighter fatalities whose injuries occurred between January 1, 1990, and December 31, 2000, in the 50 states, District of Columbia, or U.S. protectorates were included in the database and the analysis. The following definitions delineate who qualifies as a firefighter and what constitutes an on-duty fatality.

Who Is a Firefighter?

For the purpose of this study, the term *firefighter* covers all members of organized fire departments in all states, the District of Columbia, and the territories of Puerto Rico, Virgin Islands, American Samoa, Commonwealth of the Northern Mariana Islands, and Guam. Included are career and volunteer firefighters; full-time public safety officers acting as firefighters; state, territory, and federal government fire service personnel, including wildland firefighters and the military; and privately employed firefighters, including employees of contract fire departments and trained members of industrial fire brigades, whether full time or part time. It also includes contract personnel working as firefighters or assigned to work in direct support of fire service organizations.

The study includes not only local and municipal firefighters, but also seasonal and full-time employees of the U.S. Forest Service, the Bureau of Land Management, the Bureau of Indian Affairs, the Bureau of Fish and Wildlife, the National Park Service, and state wildland agencies. The definition also includes prison inmates serving on firefighting crews, firefighters employed by other governmental agencies such as the Department of Energy, military personnel performing assigned fire suppression activities, and civilian firefighters working at military installations.

What Constitutes an On-Duty Fatality?

On-duty fatalities include any injury or illness sustained while on duty that proves fatal. The term *on duty* refers to involvement in operations at the scene of an emergency, whether it is a fire or nonfire incident; being en route to or returning from an incident; performing other officially assigned duties such as training, maintenance, public education, inspection, investigations, court testimony, and fundraising; and being on call, under orders, or on standby duty, except at the individual's home or place of business.

These fatalities may occur on the fireground, in training, while responding to or returning from alarms, or while performing other duties that support fire service operations. A fatality may be caused directly by accident or injury, or it may be attributed to an occupational-related fatal illness. A common example of a fatal illness incurred on duty is a heart attack. Fatalities attributed to occupational illnesses also include a communicable disease contracted while on duty that proved fatal, where the disease could be attributed to a documented occupational exposure.

Accidents that claim the lives of on-duty firefighters are also included in the analysis, whether or not they are directly related to emergency incidents. Injuries and illnesses are included where death is considerably delayed after the original incident. When the incident and the death occur in different years, the analysis counts the fatality as having occurred in the year that the incident occurred.

It is difficult to identify an occupational illness as a causal factor in particular firefighter fatalities because of the limitations in the ability to track the exposure of firefighters to toxic hazards, the often delayed long-term effects of such exposures, and the exposures firefighters may receive while off duty.

Completeness of the Census

This analysis is based on as complete a census of on-duty firefighter fatalities as could be constructed for the 1990s. With any census, there will likely be an undercount. Despite significant research, it is not possible to claim with 100 percent assurance that all eligible firefighters are included in the analysis. In some cases, firefighters killed while on duty are not reported to the USFA nor do their survivors apply for PSOB benefits. These cases are considered to be rare, so the potential undercount should be minuscule. Thus, this census should be considered as complete as possible.

Adjustments for Unknowns

Some desired information was either unavailable or reported as "unknown" on the reports associated with each fatality. To compensate for this problem, the report cites "adjusted percentages" or percentage of valid responses or entries. The unknown items are assumed to be in the same proportion as those with known characteristics. This methodology is used by the USFA and other analysis organizations analyzing fire data. Tables and graphs note both the sample size of the valid entries and the "unknown" or unavailable entries.

FINDINGS

This section presents a statistical summary of the principal analyses undertaken in this study. Aggregate data are presented for most data elements; where possible, the information is presented graphically for clarity. Since this study sought to identify trends over time, multiyear data are also presented. In some instances, comparisons to previous firefighter fatality studies by the USFA and National Fire Protection Association (NFPA) are included.

Trend in Firefighter Fatalities

The incidence of firefighter fatalities has declined significantly over a long-term period, as shown in Figure 2. The number of fatalities annually has fluctuated from a high of 171 in 1978 to a low of 77 in 1992; overall, since 1977, firefighter fatalities have trended downward 38 percent.

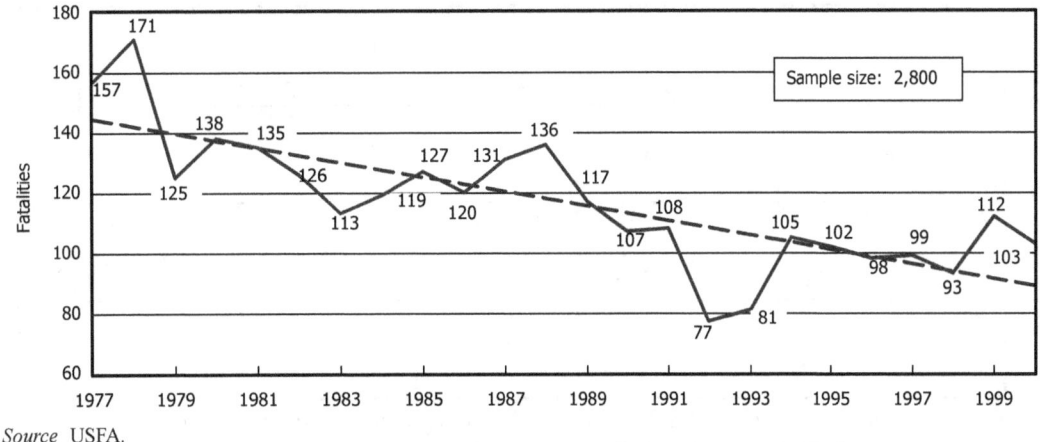

Source USFA.

Figure 2. Firefighter Fatalities (1977–2000)

During the 11-year study period, 1,085 firefighters lost their lives while on duty. (As discussed on page 4, this number differs slightly depending on the source of the data and the criteria used to define an on-duty fatality.) Despite the overall decline in firefighter fatalities since 1977 and a sharp decline in firefighter fatalities between 1991 and 1992, the incidence of on-duty firefighter fatalities has trended upward 7 percent since 1990.

Type of Incident

Figure 3 illustrates the types of incidents that results in firefighter fatalities during the study period. Not all firefighters were involved in an emergency incident at the time of their fatal injury (e.g., physical fitness, administrative duties). The two leading types of incident (structural fire/explosion and wildland/brush fire) account for 67 percent of firefighter fatalities.

Figure 4 shows the rate of firefighter fatalities per 100,000 reported fire incidents. Despite wide fluctuations, the overall trend in this rate declined 15 percent between 1983 and 2000. Over the study period itself, however, fire-related firefighter fatalities per 100,000

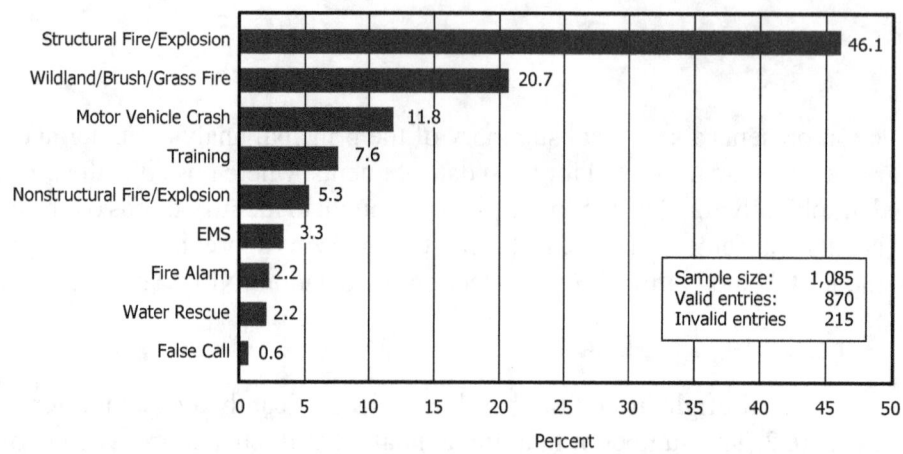

Figure 3. Types of Incident Resulting in Fatalities (1990–2000)

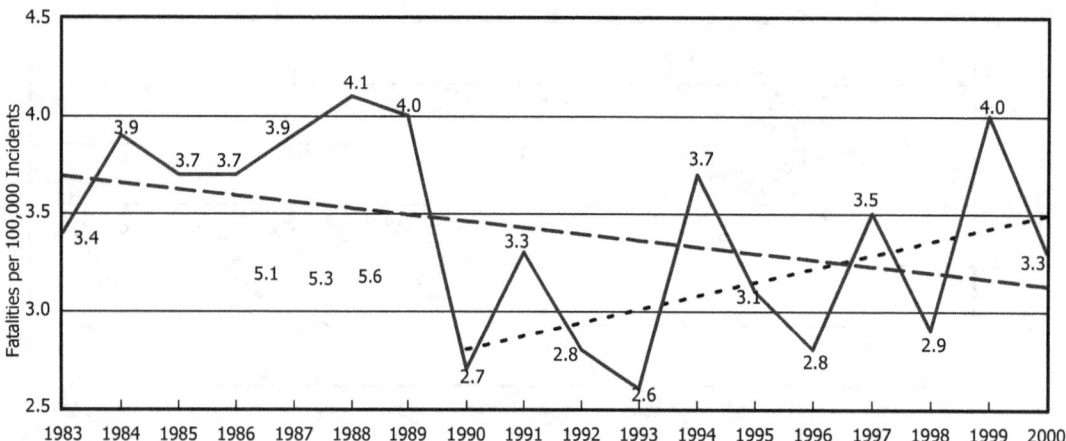

Note: These data include only firefighters who were reportedly engaged in response to/return from an incident, fire extinguishment/incident neutralization, or suppression support activities at the time of their deaths.

Source USFA and NFPA.

Figure 4. Fire-Incident-Related Firefighter Fatalities per 100,000 Incidents (1977–2000)

reported incidents has risen approximately 25 percent, with 1999 having the second highest rate since 1988. By sharp constrast, the trend in fire incidence declined 15 percent over the study period [Ref. 3].

Affiliation of Firefighter Fatalities and Type of Agency

Figure 5 illustrates the distribution of firefighter fatalities by affiliation. The majority of firefighter fatalities, 57 percent, were members of local or municipal volunteer fire agencies (including combination departments, which are comprised of both career and volunteer personnel). Full-time career personnel account for 33 percent of firefighter fatalities; however, they comprise only 26 percent of the American fire service. Therefore, although numerically more volunteer firefighters are killed than career personnel, career personnel are killed at a

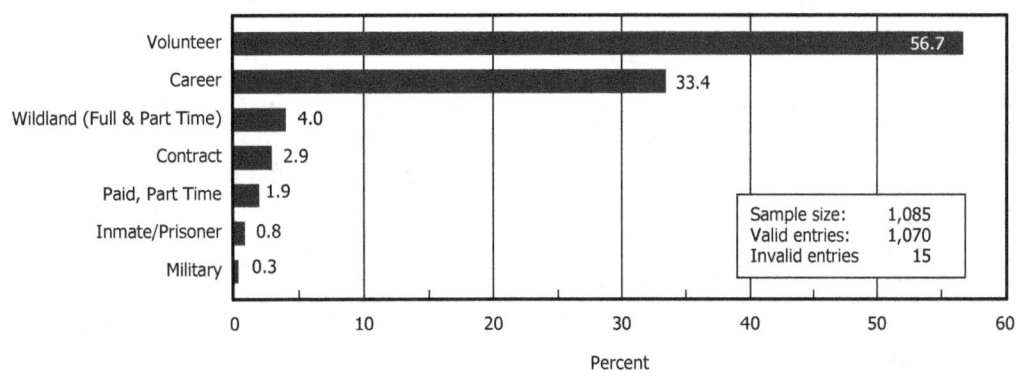

Figure 5. Firefighter Fatalities by Affiliation (1990–2000)

rate disproportionate to their representation in the fire service. In part, this disproportion may be attributable to disparities in the call volume and types of calls to which career personnel typically respond. Further, career departments protect more of the U.S. population than do volunteer agencies (59 percent vs. 41 percent), which also affects the volume and types of calls to which personnel respond [Ref. 4].

Wildland firefighters (full- and part-time wildland firefighters, contract personnel, and prisoners), account for 8 percent of firefighter fatalities during the study period.

Figure 6 illustrates the distribution of career and volunteer firefighter fatalities during the study period. Although there has been some fluctuation from year to year, the general distribution has not changed significantly over time with about one-third of fatalities volunteer and approximately 55 percent career. The remainder of personnel were part-time, wildland,

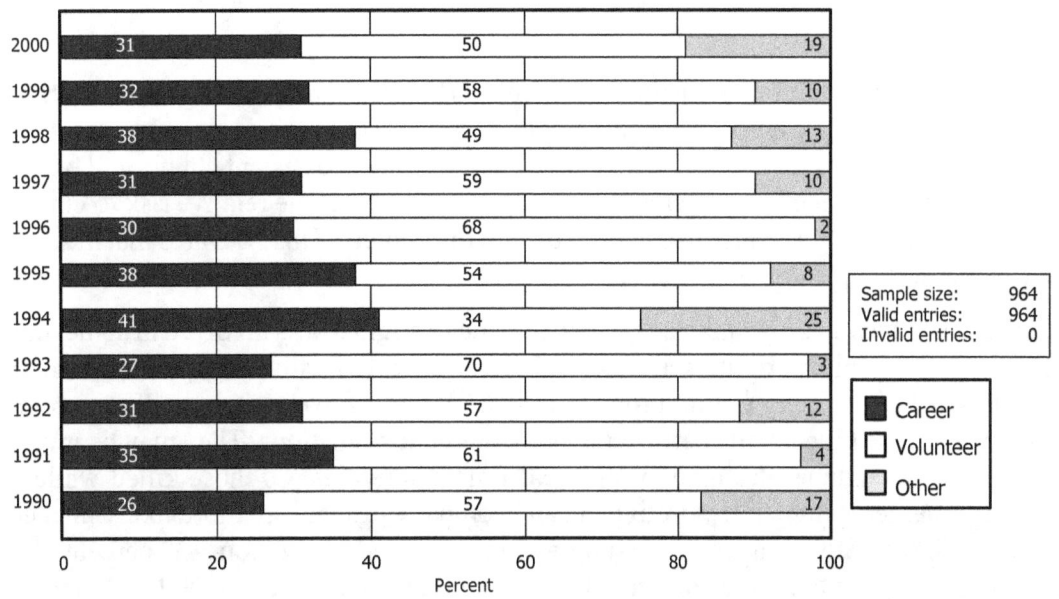

Figure 6. Career–Volunteer Distribution (1990–2000)

15

military, or contract personnel. (In 1994, the distribution was significantly altered by the deaths of 14 wildland firefighters at Storm King Mountain.)

Nationally, 73 percent of fire department agencies are all volunteer, 21 percent are combination, and 6 percent are all career. [Ref. 4]. As with individual affiliation, career fire departments experience a disproportionate number of firefighter fatalities. Figure 7 illustrates the types of agencies that have experienced firefighter fatalities. Fifty-six percent of firefighter fatalities were associated with volunteer organizations, 9 percent with combination departments, and 28 percent with career departments. (Wildland agencies are not specified separately here; rather, such agencies are distributed among career and federal agencies, as well as private contractors.)

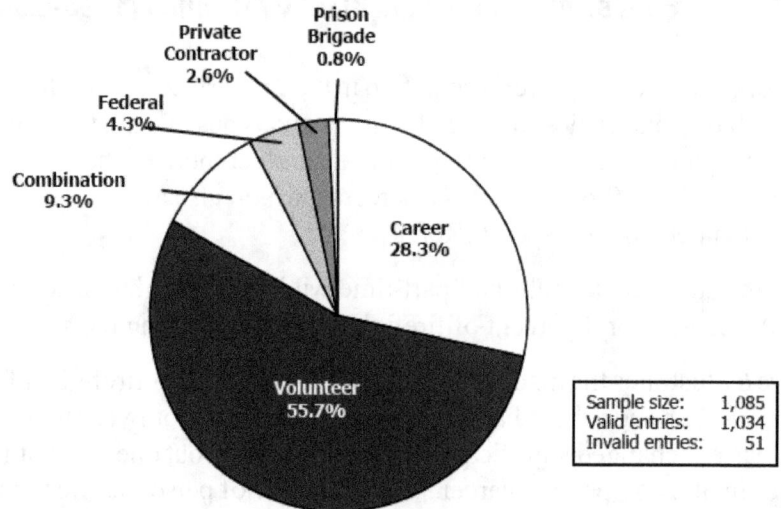

Figure 7. Type of Agency with Firefighter Fatality (1990–2000)

Rank of Firefighter

The majority of firefighter fatalities (60 percent) held the rank of firefighter at the time of their death (Figure 8). Company officers (captains, lieutenants, and sergeants) account for 15 percent of fatalities; chief officers (fire chiefs, deputy chiefs, assistant chiefs, and battalion chiefs) account for another 13 percent of firefighter fatalities. The relative risk to chief and company officers is greater than that faced by line firefighters. This is an area that merits further research.

Figure 9 illustrates, by rank, the type of duty the firefighter was involved in at the time of injury. Although all firefighters are more likely to be killed while engaged in emergency activities, battalion chiefs, recruit/probationary firefighters, fire marshals, and inspectors are more likely to be killed while performing non-emergency functions. They may be involved in administrative duties, training, or physical fitness activities. Of those killed while performing emergency functions, firefighters and company officers are more likely than chief officers to be actually engaged in fire suppression or incident mitigation—32 percent of firefighters, 45 percent of lieutenants, and 33 percent of captains as compared to 13 percent of assistant/deputy chiefs and 17 percent of fire chiefs.

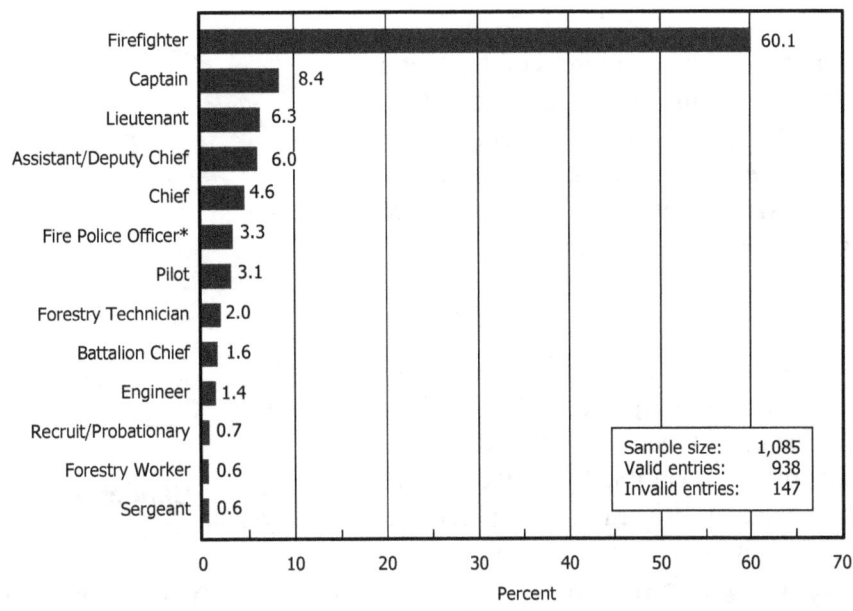

*Several states, including Pennsylvania, New York, and New Jersey, have enacted legislation that allows volunteer fire departments to assign specially trained firefighters limited law eenforcement powers while operating on the scene of an emergency incident.

Figure 8. Firefighter Fatalities by Rank (1990–2000)

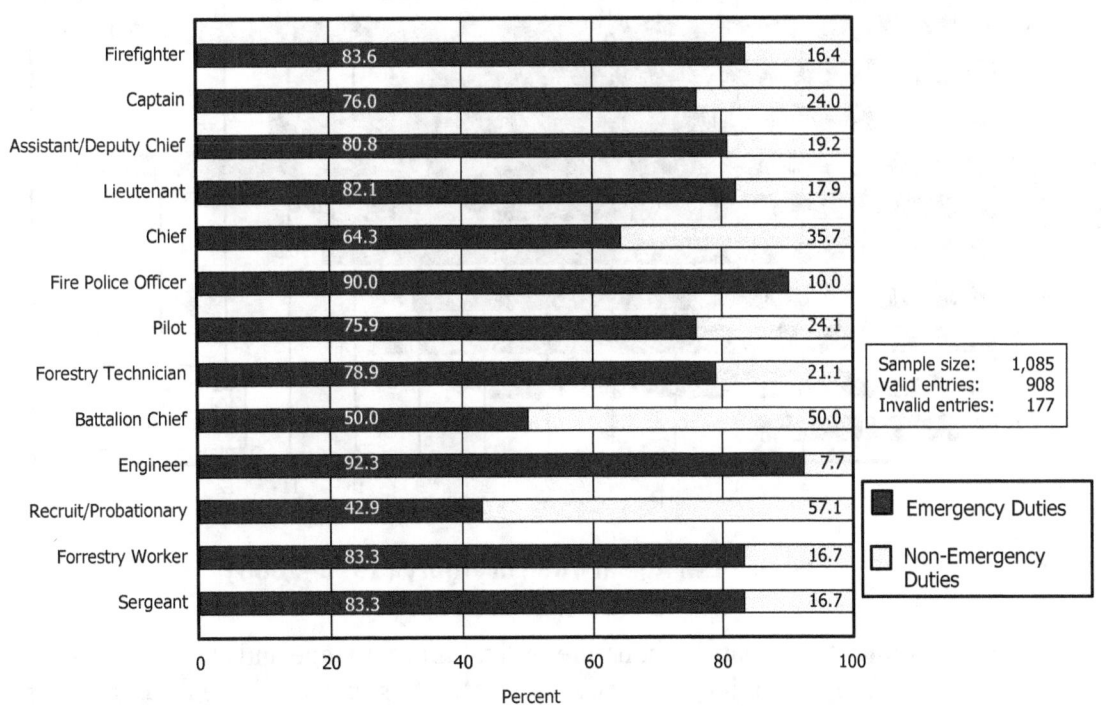

Figure 9. Rank of Firefighter vs. Type of Duty at Time of Fatal Injury (1990–2000)

17

Gender

Consistent with the demographics of the American fire service, the vast majority of firefighter fatalities are male; only 3 percent are female (Figure 10). As more women become career and volunteer firefighters, females will likely comprise a higher proportion of firefighters killed while on duty.

Age

Figure 11 shows the age distribution of firefighter fatalities. Fifty-nine percent of firefighters were over the age of 40 when they were killed, and one-third were over 50. Nationally, however, firefighters over the age of 40 comprise 46 percent of the fire service, with those over 50 accounting for only 16 percent of firefighters [Ref. 4]. Older firefighters possess a wealth of invaluable knowledge and experience, but they are killed while on duty at a rate disproportionately high to their representation in the fire service. Also, these older firefighters tend to be affiliated with volunteer agencies. About 40 percent of volunteer firefighters are over the age of 50, compared to only 25 percent of career firefighters.

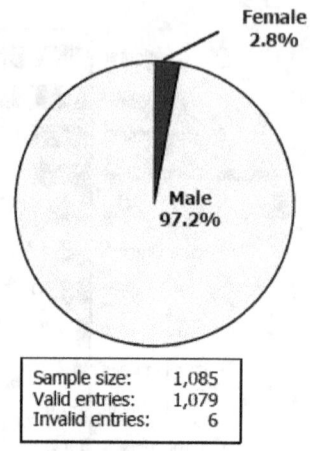

Sample size:	1,085
Valid entries:	1,079
Invalid entries:	6

Figure 10. Gender of Firefighter Fatalities (1990–2000)

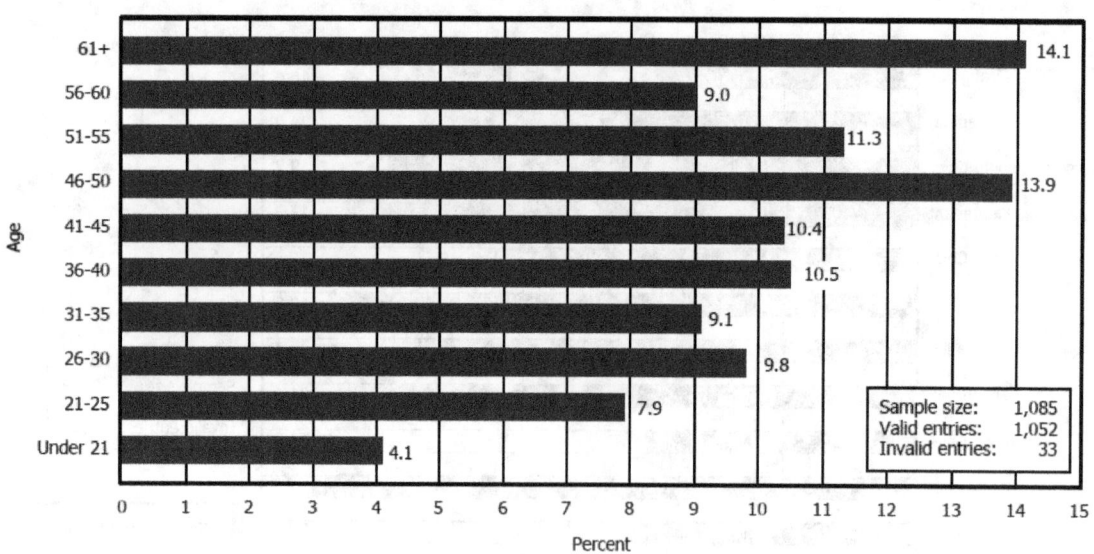

Sample size:	1,085
Valid entries:	1,052
Invalid entries:	33

Figure 11. Age at Time of Injury (1990–2000)

Table 1 illustrates the variations in nature of fatal injury by age and shows the distinct difference in nature of fatal injury after age 35. Firefighters under the age of 35 are more likely to be killed by traumatic injuries, such as internal trauma and burns, than they are to die of medical causes (e.g., heart attack, stroke). After age 35, the proportion of deaths from traumatic injuries decreases, and the proportion of deaths from heart attacks and other medical causes rises steadily. An interesting pattern emerges for firefighters between the ages of 36

Table 1. Leading Nature of Fatal Injury by Age (1990–2000)

Age	Burns/ Asphyxiation	Medical	Trauma	Other	Total
Under 21	23%	5%	70%	2%	100%
21–25	33%	6%	52%	10%	100%
26–30	46%	11%	43%	1%	100%
31–35	30%	17%	48%	5%	100%
36–40	**30%**	**34%**	**33%**	**4%**	**100%**
41–45	23%	47%	23%	6%	100%
46–50	10%	59%	22%	8%	100%
51–55	9%	66%	20%	5%	100%
56–60	3%	76%	19%	2%	100%
61+	4%	78%	16%	2%	100%

Sample size: 1,085
Valid entries: 1,052
Invalid entries: 33

and 40. In this age group, medical causes, burns/asphyxiation, and trauma each account for approximately one-third of fatalities.

Figure 12 shows the variance in type of duty by age of the firefighter. Younger firefighters were more likely than older firefighters to be engaged in incident mitigation or training than response or suppression support at the time of their fatal injury.

Table 2 illustrates, by age group, whether a firefighter was engaged in emergency or non-emergency duties at the time of his or her death. Firefighters of all ages were more likely to be engaged in emergency duties at the time of their death.

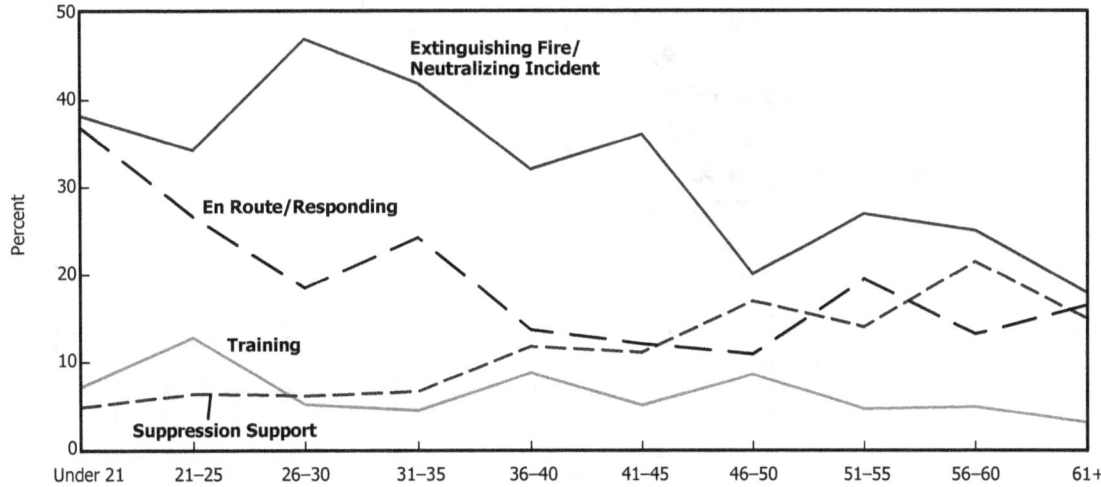

Figure 12. Type of Duty by Age of Firefighter (1990–2000)

Table 2. Emergency vs. Non-Emergency Duties by Age of Firefighter (1990–2000)

Age	Emergency Duties	Non-Emergency Duties
Under 21	81%	19%
21–25	81%	19%
26–30	83%	17%
31–35	84%	16%
36–40	81%	19%
41–45	78%	22%
46–50	68%	32%
51–55	78%	22%
56–60	77%	23%
61+	86%	14%

Type of Duty

Figure 13 illustrates the types of duties firefighters were engaged in at the time of their fatal injury. Prior to 1990, the largest share of fatal firefighter injuries have happened either on the fireground[3] or en route to an incident [Ref. 5]. Continuing this trend, during the study period the leading activity at the time of injury was extinguishing fire/neutralizing the incident (30 percent), followed by responding to the scene (18 percent) and suppression support

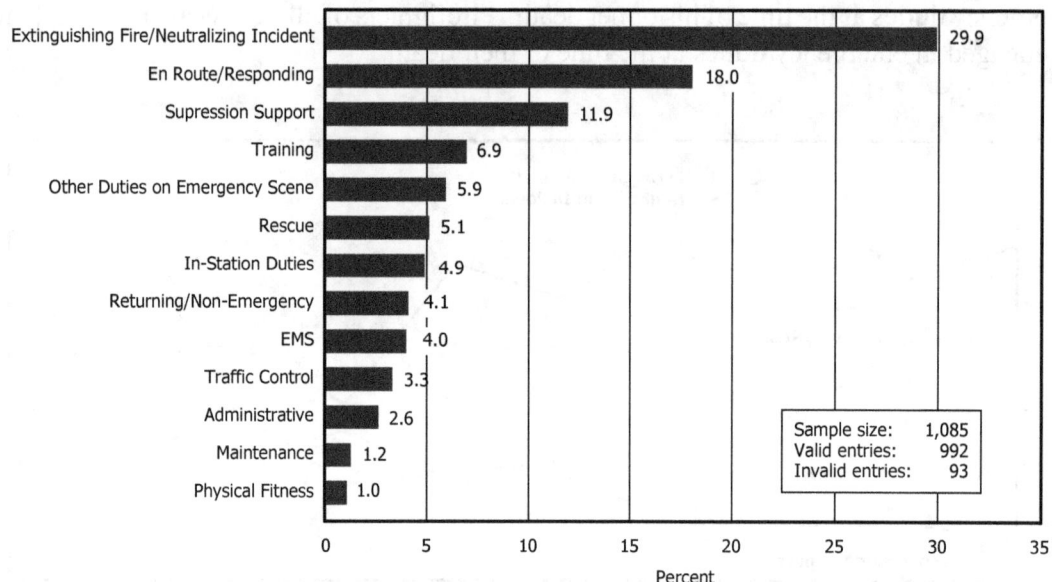

Figure 13. Type of Duty at Time of Fatal Injury (1990–2000)

---

[3]For this analysis, it was not possible to isolate firefighter fatalities on the fireground from those that occurred elsewhere. Instead, the analysis considered whether the firefighter was engaged in emergency or non-emergency duties at the time of his or her fatal injury. These categories are defined later in this section.

(12 percent), which includes forcible entry, ventilation, salvage and overhaul, and raising ground ladders.

Emergency Medical Services (EMS)—In many fire departments, EMS calls account for between 50 and 80 percent of emergency call volume. Yet such incidents result in only 3 percent of firefighter fatalities. Moreover, 4 percent of fatally injured firefighters were performing EMS duties at the time of their deaths. That is, some firefighters were performing EMS functions while operating on a non-EMS call/incident (e.g., a motor vehicle collision). Trauma accounts for the deaths of 50 percent of firefighters who were involved in EMS operations at the time of their fatal injury; another 38 percent involved in EMS operations died from heart attacks. The higher ratio of deaths due to trauma may be a result of collisions while transporting patients, a category heading not included in the database.

Emergency vs. Non-Emergency—Figure 14 groups the duties in Figure 13 in two categories: emergency and non-emergency. Emergency duties include extinguishing/neutralizing the incident, responding, suppression support, rescue, EMS, and other duties at the scene. Non-emergency duties are training, physical fitness, administrative, maintenance, and other in-station activities. Most firefighter fatalities (79 percent) are involved in the mitigation of an emergency incident or response to an emergency at the time of their injury; 21 percent do not involve emergency duties.

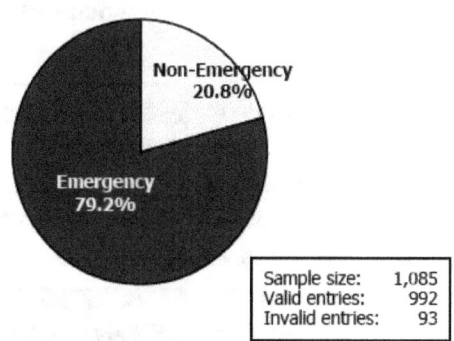

Figure 14. Emergency vs. Non-Emergency Duty (1990–2000)

Training—Since 1990, approximately 6 percent of firefighter fatalities occurred during training activities, a larger proportion than in the previous decade.[4] Over time, the leading type of training activity resulting in fatalities has remained physical fitness, followed by apparatus/equipment drills and live-fire exercises. During training activities, the leading nature of fatal injury is heart attack (54 percent), followed by trauma (31 percent).

Figure 15 groups training fatalities by type of training activity. The high incidence of deaths during physical fitness is troubling. The dangers of live-fire exercises are well known and awareness is generally high, but it may be a surprise that more firefighters are killed during equipment/apparatus drills than in live-fire training.

Motor Vehicle Collisions

Since 1984, motor vehicle collisions (MVCs) have accounted for between 20 and 25 percent of firefighter fatalities annually [Ref. 7]. During the study period, MVCs accounted for 22 percent of firefighter fatalities. One quarter of firefighters who died in MVCs were killed in private/personally owned vehicles (POVs) (Figure 16). Following POVs, the apparatus most often involved in fatal collisions were tankers, engines/pumpers, and airplanes. More

---

[4]Between 1978 and 1987, however, training accounted for less than 5 percent of firefighter fatalities [Ref. 6].

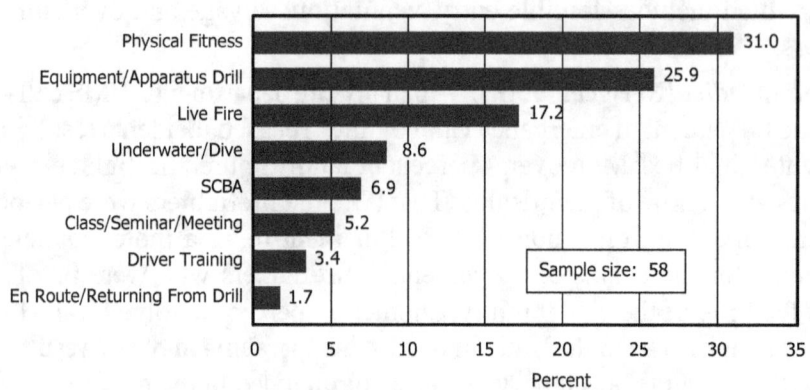

Figure 15. Leading Types of Training Activities Associated With Fatalities (1990–2000)

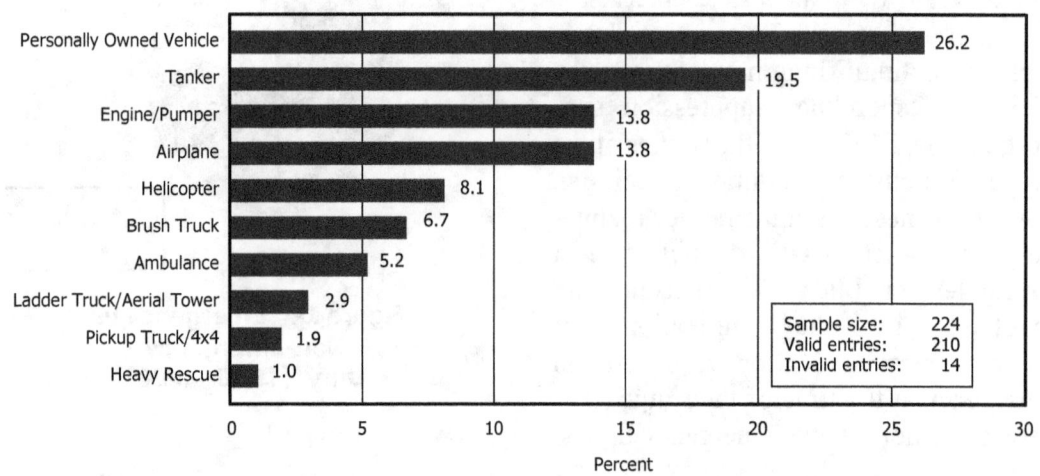

Figure 16. Type of Vehicle Involved in Collision (1990–2000)

fatalities occur in tanker collisions than in engines and ladders combined. Approximately 27 percent of fatalities in MVCs were ejected from the vehicle at the time of the collision; only 21 percent of these firefighters were reportedly wearing their seatbelts prior to the collision.

Most volunteer departments do not require personnel to stand by in the fire station. Rather, members are allowed to respond directly to incidents from their homes or work-places, often in their POVs. As a result, volunteers are more likely than career firefighters to be killed in POV collisions. Moreover, they are more likely to be involved in collisions involving tankers, which are predominantly used in rural areas without hydrants or other readily available sources of water. Such areas are almost exclusively protected by volunteer fire departments.

Tankers that are overloaded or whose tanks lack proper baffling can be unstable, making them more difficult to control. The shifting of water in the tank, even at low speeds, can dramatically affect the ability of the apparatus operator to control the vehicle. Water weighs approximately 8.4 pounds per gallon. An average tanker holds between 2,000 and 3,000 gallons of water, which adds between 16,800 and

25,200 pounds to the weight of vehicle. The incidence of tanker collisions is an area that merits further investigation.

Figure 17 shows the locations of firefighters in the vehicle prior to the fatal collision. The majority were driving the apparatus prior to their death, as would be expected for incidents involving POVs or tankers, which often respond to calls with only a driver.

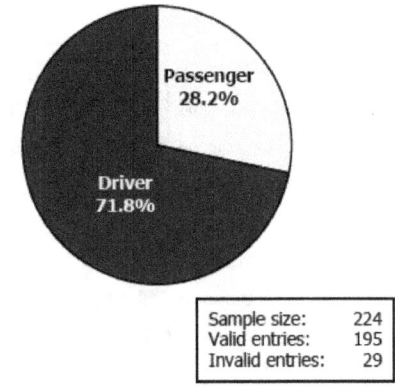

Sample size:	224
Valid entries:	195
Invalid entries:	29

Immediate Cause and Nature of Fatal Injury

Figure 17. Decedent's Location in Vehicle (1990–2000)

The causes of fatal firefighter injuries are illustrated in Figure 18. The leading cause, overexertion/strain, is consistent with the high incidence of deaths from heart attacks (discussed later in this chapter) and accounts for nearly half of firefighter deaths. Other leading causes of firefighter injuries are being trapped, caught, or lost in a structure; fire apparatus collisions; and being struck by an object (e.g., vehicle, apparatus, falling debris in a structure).

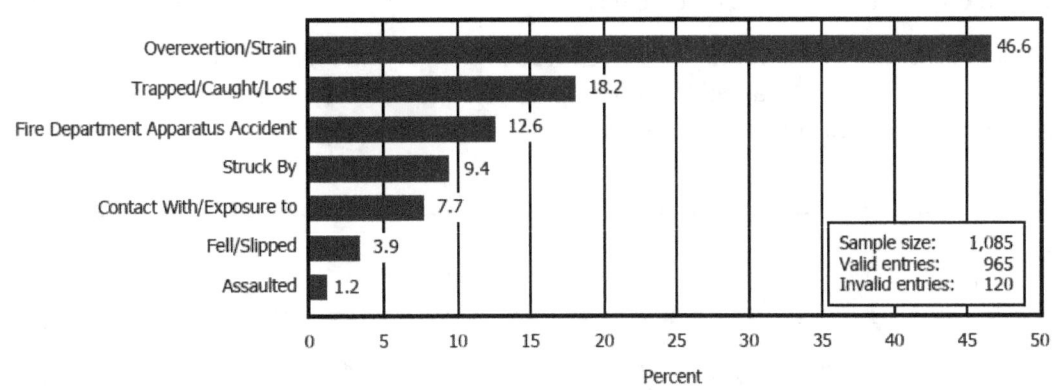

Figure 18. Immediate Cause of Fatal Injury (1990–2000)

These are only the immediate causes of injury; there is almost always a chain of events that leads to fatalities. For example, the *reason* a firefighter gets trapped and dies may be because of a lack of adequate situational awareness by the incident commander, a dangerously weakened structure that went undetected, the lack of a way to find the trapped firefighter quickly enough, a shift in wind conditions on a wildland fire, or poor judgment on risk taking.

The leading nature of fatal injuries to firefighters is heart attack (44 percent), as shown in Figure 19. Trauma, including internal and head injuries, is the second leading type of fatal injury at 27 percent. Asphyxia and burns combined account for 20 percent of fatalities. Thus, more firefighters die from trauma than from asphyxiation and burns combined.

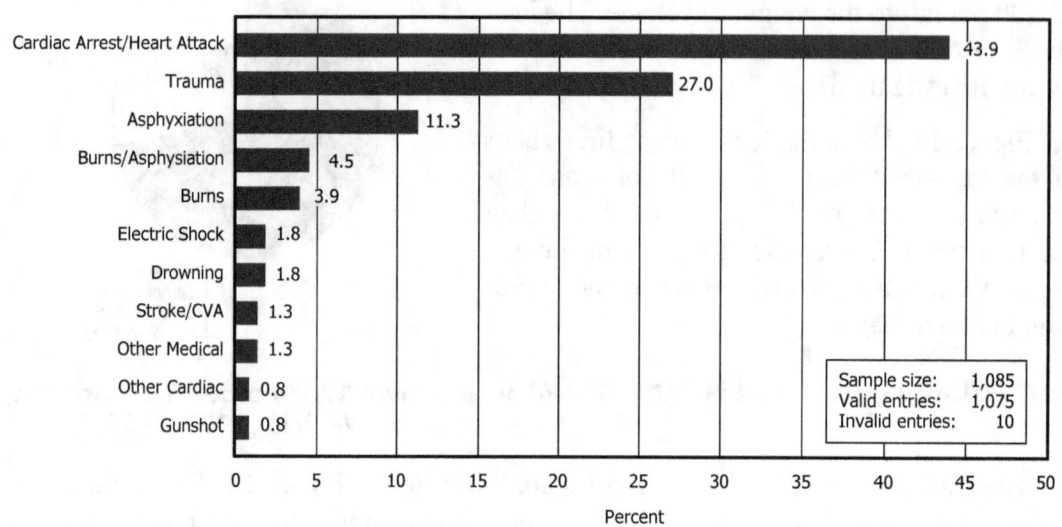

Figure 19. Nature of Fatal Injury (1990–2000)

Figure 20 shows the trend in percent of deaths due to heart attack from 1984–2000. Despite fluctuations, the trend in the proportion of firefighter fatalities from heart attacks has remained constant over the past 16 years.

Where reported by the family or discovered at autopsy, the most common pre-existing condition found for heart attack fatalities was arteriosclerosis,[5] followed by prior heart attack(s) and hypertension.

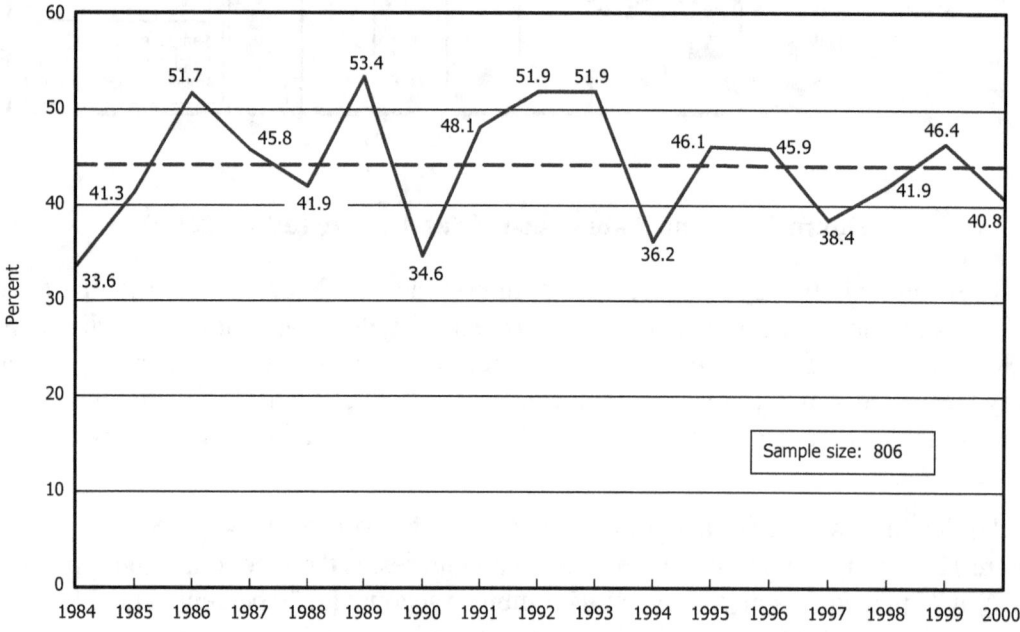

Source pre 1990 data and USFA *Analysis Report on Firefighter Fatalities*, U.S. Fire Administration, FEMA, August 1994.

Figure 20. Percent of Heart Attack Deaths by Year (1984–2000)

---

[5]Arteriosclerosis is the progressive hardening of the arteries over time.

24

Firefighters are more likely to suffer a heart attack in the course of performing suppression support duties on the fireground, while in the fire station, or during training exercises. In contrast, deaths due to traumatic injuries are more likely to occur while mitigating or responding to an incident (Figure 21).

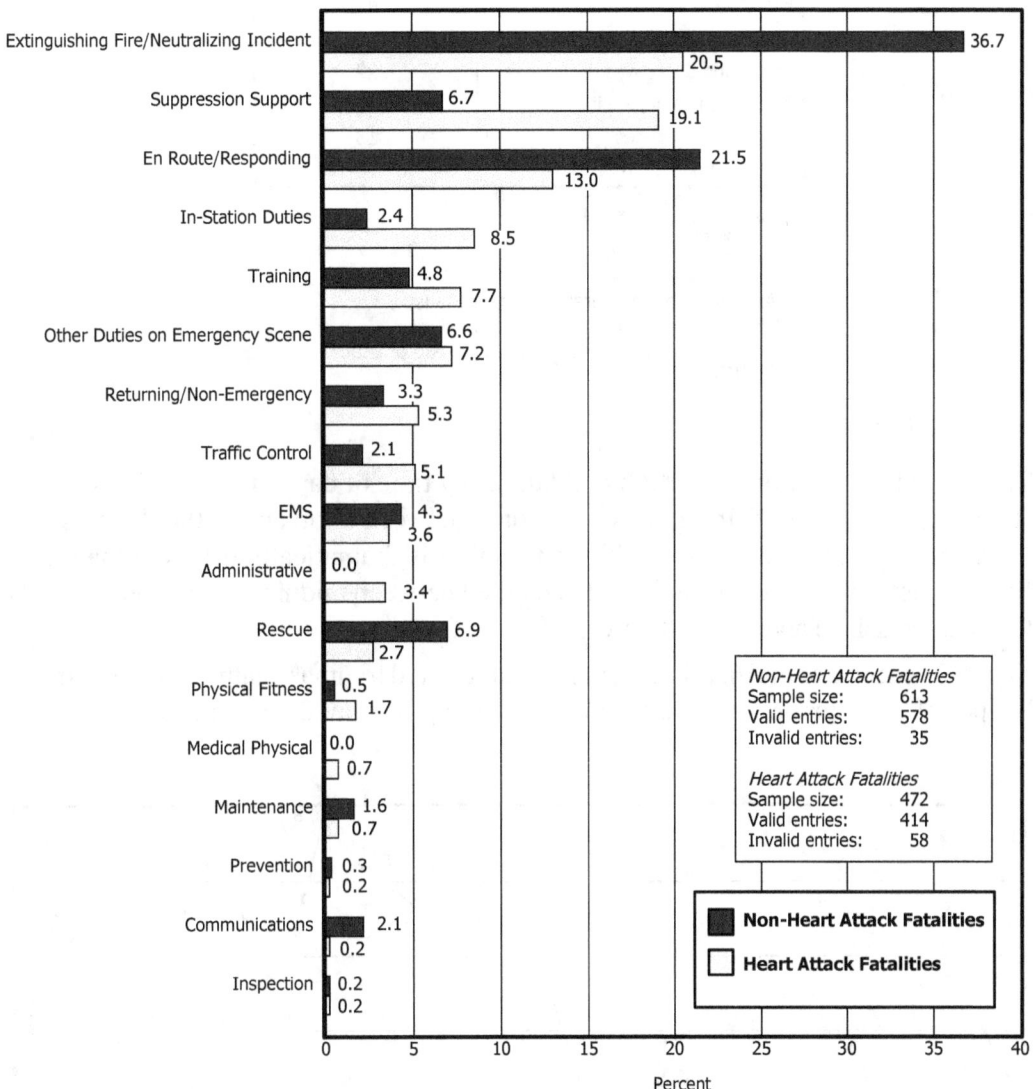

**Figure 21. Comparison of Type of Duty for Heart Attack
With Non-Heart Attack Fatalities (1990–2000)**

Comparison to Other Fatality Rates

Table 3 compares heart attack rates for firefighters with other occupational categories that require significant physical labor (e.g., construction) or have similar periods of downtime followed by intense activity (e.g., police officers). Although the term used here is *occupation,* the firefighter category includes all categories of affiliation (career, volunteer, wildland, etc.). Firefighters, as a group, are more likely than other American workers to die of a heart attack while on duty.

Table 3. Comparison of Heart Attack Fatalities by Occupation (1990–2000)

Occupation*	Percent of Deaths Due to Heart Attacks
Firefighters	44**
Guards (including supervisors)	25
Police and Detectives	22
All Occupational Fatalities	15
Construction Trades	13
Construction Laborers	10

Sample size:	1,085
Valid entries:	1,075
Invalid entries:	10

* As defined by the Bureau of Labor Statistics.
** Based on data collected for this analysis.
Source: Excluding firefighter data, Bureau of Labor Statistics.

Time of Injury

The overall distribution of firefighter fatalities by time of day is illustrated in Figure 22 and Table 4. Nearly half of firefighters are injured between noon and 2000; the most common time of fatal injury was 1600. This distribution is dramatically different than that of civilian fire victims, who are most likely to be killed between midnight and 0600, when they are likely to be asleep and unable to escape from a fire.

Figure 23 compares the distribution of heart attack and trauma[6] deaths by hour of the day. Both types of fatalities are more likely to occur during the day than at night.

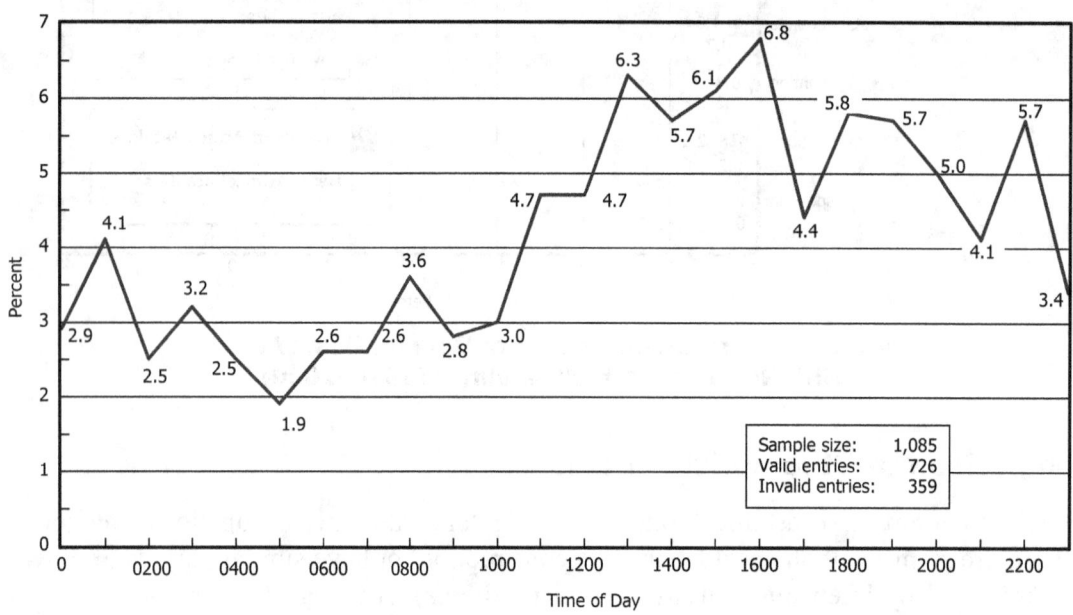

Figure 22. Percent of Injuries by Time of Day (1990–2000)

---

[6]For this particular analysis, only deaths coded as trauma (internal, head, etc.) were considered.

Table 4. Time of Injury (1990–2000)

Time of Day	Percent of Fatalities
0000–0359	13
0400–0759	10
0800–1159	14
1200–1559	23
1600–1959	23
2000–2359	18

Sample size: 1,085
Valid entries: 726
Invalid entries: 359

Sample size: 762
Valid entries: 489
Invalid entries: 273

Figure 23. Percent of Trauma vs. Heart Attack Deaths by Hour of the Day (1990–2000)

Fixed Property Use

Despite recent firefighter fatality incidents involving abandoned structures (e.g., Worcester, Massachusetts, 1999) and highrises (e.g., New York City, New York, 1998), each of these properties account for only approximately 3 percent of fatalities. Twenty-eight percent of firefighter fatalities are killed during incidents on residential properties, as shown in Figure 24. For civilian fire casualties, between 75 and 85 percent occur on residential properties. The prominence of outdoor properties is a result of the high incidence of wildland fires. Further research is needed to determine the relationship between the dangers posed by particular properties and the incidence of firefighter fatalities.

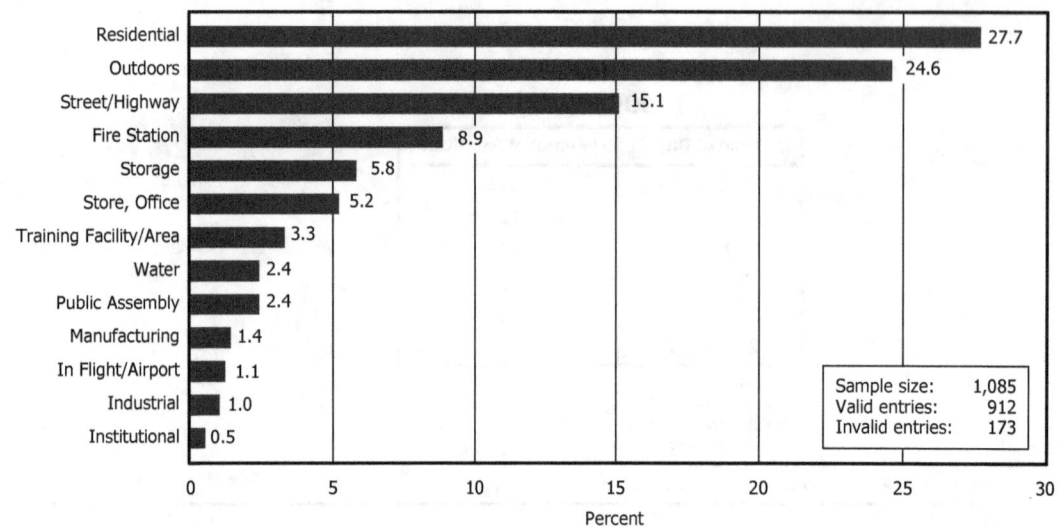

Figure 24. Fixed Property Use of Incident Where Injury Occurred (1990–2000)

Of those firefighters killed while en route to an incident (179 cases), 85 percent were volunteers. However, for firefighters killed performing in-station duties, 69 percent were career personnel; the majority of those deaths were the result of heart attacks. These variations can be attributed to differences between career and volunteer agencies. Generally, unless they are on a call or other fire department business, career personnel are required to be in the fire station for the duration of their shift, which is generally between 10 and 24 hours long. Volunteers, on the other hand, are not usually required to stand by at the station, so they can respond from their homes or places of work. As a result, career personnel are more likely than volunteers to die in a fire station.

For other property types, the ratio of volunteer to career firefighter fatalities is approximately equal. As would be expected, wildland, contract, and inmate firefighters who are killed are outdoors (including those in aerial missions).

Cause of Fire[7]

As illustrated in Figure 25, in firefighter fatality incidents where a fire is involved, the most common fire cause is incendiary/suspicious (arson) at 37 percent. Other leading causes of fatal fires include electrical distribution, natural, and open flame. For civilian fire casualties, the leading causes of fatal fires are smoking, arson, and heating [Ref. 8].

Explanations for the differences in cause associated with firefighter vs. civilian fire fatalities are varied. For example, in arson fires, the use of accelerants contributes to rapid fire spread and growth and can lead to a fire that causes substantial structural damage in a short amount of time. Moreover, since they may not be detected and reported for some time after being set, arson fires may have the opportunity to advance more than other types of fires before the arrival of the fire department. Additionally, wildland and other outdoor fires are

---

[7]Cause categories are based on the Priority Cause Grouping Code used in NFIRS, but were assigned based on incident reports submitted by fire departments or as reported in the media.

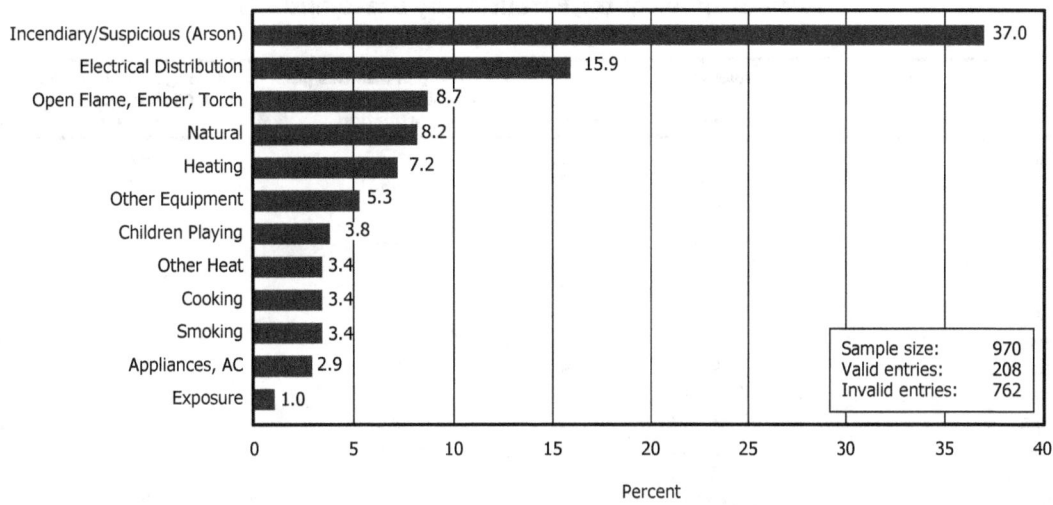

Figure 25. Cause of Fire, If a Fire Incident (1990–2000)

more likely to be caused by open flame or natural causes than are residential structure fires (where the majority of civilian fatalities occur). The relationship between fire cause and the incidence of firefighter fatalities is an area that merits further research.

Geographical Distribution of Firefighter Fatalities

Generally, firefighter fatalities are associated with the state where their agency is located. Table 5 illustrates the incidence of firefighter fatalities by state and compares the incidence of fatalities per capita by the state of the firefighter's affiliation versus the state where he or she was actually injured. This difference is particularly notable for Colorado where 14 wildland firefighters were killed in 1994. However, of those 14, most were affiliated with agencies based in other states, including Oregon, South Carolina, and Idaho.

Figure 26 graphically illustrates the incidence of firefighter fatalities nationally.

Figure 27 illustrates the firefighter fatalities involved in wildland, MVC, and structural incidents per capita. Figure 28 illustrates the distribution of fatalities affiliated as career, volunteer, or wildland firefighters per capita. Firefighter fatalities in structural incidents are more common in the densely populated East; firefighter fatalities in wildland incidents predominantly occur in the West. MVC fatalities have no defined pattern.

Volunteer firefighter fatalities are fairly well distributed throughout the Midwest and eastern United States. A similar distribution is seen in career firefighter fatalities. As would be expected, wildland firefighter fatalities (including contract employees and prisoners) are mostly concentrated in the western states.

Large population states generally have more fatalities than those with smaller populations. However, Florida has a relatively high population and a low rate of firefighter fatalities per 10 million population. There are other surprises. New York and Pennsylvania have many more fatalities than California and Texas, even though California and Texas have higher populations. In fact, Texas and New York have comparable average populations, yet New York has more than twice as many fatalities as Texas. Also surprising is that the state with the

Table 5. Firefighter Fatalities by State and Per 10 Million Population (1990–2000)*

State	Average 1990–2000 Population	Total Fatalities	Average Fatalities per 10 Million Population by Affiliation	Average Fatalities per 10 Million Population by Incident
Alabama	4,243,844	20	47.1	47.1
Alaska	588,488	2	34.0	34.0
Arizona	4,397,930	15	34.1	34.1
Arkansas	2,512,063	14	55.7	55.7
California	31,815,835	75	23.6	21.7
Colorado	3,797,828	7	18.4	52.7
Connecticut	3,346,341	20	59.8	59.8
Delaware	724,884	3	41.4	41.4
District of Columbia	695,250	4	57.5	57.5
Florida	14,460,152	22	15.2	15.2
Georgia	7,332,335	21	28.6	28.6
Hawaii	1,159,883	4	34.5	34.5
Idaho	1,150,351	10	86.9	69.5
Illinois	11,924,948	32	26.8	36.9
Indiana	5,812,322	44	75.7	55.1
Iowa	2,851,540	14	49.1	49.1
Kansas	2,582,996	14	54.2	54.2
Kentucky	3,863,533	25	64.7	64.7
Louisiana	4,344,475	12	27.6	25.3
Maine	1,251,426	7	55.9	55.9
Maryland	5,038,977	29	57.6	57.6
Massachusetts	6,182,761	27	43.7	43.7
Michigan	9,616,871	22	22.9	22.9
Minnesota	4,647,289	6	12.9	12.9
Mississippi	2,708,937	18	66.4	66.4
Missouri	5,356,142	24	44.8	44.8
Montana	850,630	4	47.0	58.8
Nebraska	1,644,824	6	36.5	36.5
Nevada	1,600,045	5	31.2	37.5
New Hampshire	1,172,519	4	34.1	34.1
New Jersey	8,072,269	37	45.8	45.8
New Mexico	1,667,058	12	72.0	66.0
New York	18,483,456	126	68.2	68.2
North Carolina	7,338,975	25	34.1	34.1
North Dakota	640,500	3	46.8	46.8
Ohio	11,100,128	36	32.4	32.4
Oklahoma	3,298,120	21	63.7	63.7
Oregon	3,131,860	19	60.7	28.7
Pennsylvania	12,081,349	97	80.3	81.9
Rhode Island	1,025,892	4	39.0	39.0
South Carolina	3,749,358	19	50.7	48.0
South Dakota	725,424	2	27.6	27.6
Tennessee	5,283,234	24	45.4	45.4
Texas	18,919,165	60	31.7	31.7
Utah	1,978,010	6	30.3	30.3
Vermont	585,793	8	136.6	136.6
Virginia	6,632,937	22	33.2	33.2
Washington	5,380,407	15	27.9	22.3
West Virginia	1,800,911	16	88.8	88.8
Wisconsin	5,127,722	14	27.3	27.3
Wyoming	473,685	6	126.7	84.4
Puerto Rico	3,665,324	3	8.2	8.2

Affiliation			*Incident*			*Calculations are based on the average number of fatalities per average population.
Sample size:	1,085		Sample size:	1,085		
Valid entries:	1,085		Valid entries:	1,076		
Invalid entries:	0		Invalid entries:	9		

Figure 26. Map of Firefighter Fatalities, by State of Affiliation (1990–2000)

highest per capita rate (per 10 million population) of firefighter fatalities is Vermont. Given Vermont's relatively low population, however, even one firefighter fatality would result in a rate of 17 firefighter fatalities per 10 million population. As such, Vermont's seemingly high rate of firefighter fatalities is not conclusive. Rather, it is possible that the rate is abnormally high due to the analytic metric being used. Using comparisons based on other data (e.g., the population of firefighters in Vermont and the composition of the state's fire service) could result in a very different rate. The state with the lowest rate of firefighter fatalities per 10 million population is Minnesota. Again, however, the limitations of this analytic metric make it difficult to determine the validity of this estimate. Further study is necessary to overcome the issues created by the use of population data.

To explore some of the reasons for the differences in the patterns of firefighter fatalities by state, the tables in this section compare the experience of firefighter fatalities in various states throughout the country. The experiences of various states are highlighted to determine possible reasons for their dissimilar experiences with regard to firefighter fatalities. However, much additional work is required to definitively determine the reasons for the different patterns associated with firefighter fatalities by state.

The states compared in this section were chosen for their geographic location, population, and climate. Minnesota, Wisconsin, and Michigan are viewed as a single entity to bring their total population in line with the comparison group. These states have similar climates and are all located in the upper Midwest. The hypothesis being that since population alone does not explain variances between states, other factors must exert varying degrees of influence. These factors include climate, housing stock, life/safety codes, age of firefighters, types of fires to which the fire department responds, and distribution of personnel (career,

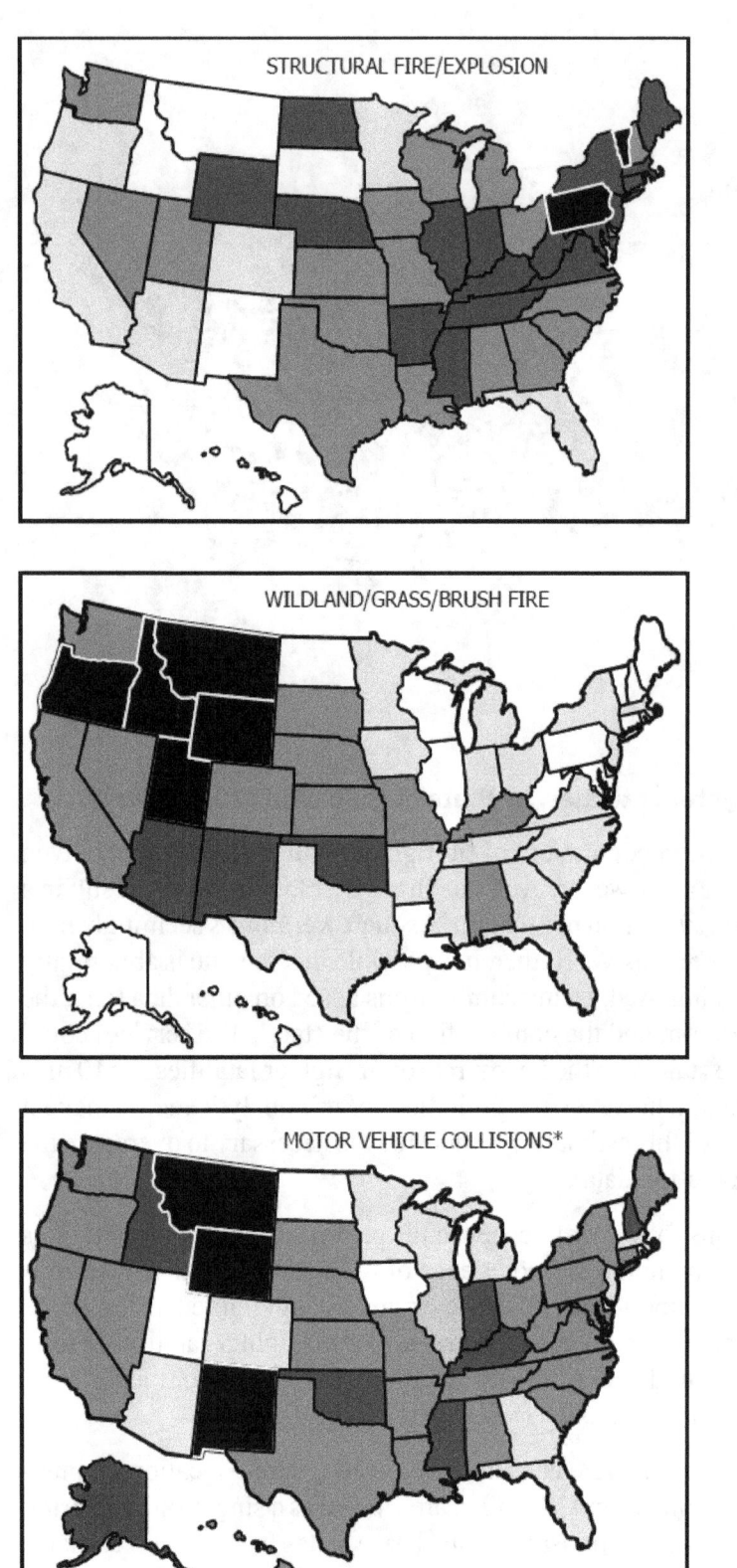

*Reflects incidents where the firefighter fatality
was involved in an MVC.

Figure 27. Firefighter Fatalities, by Affiliation Per 10 Million Population

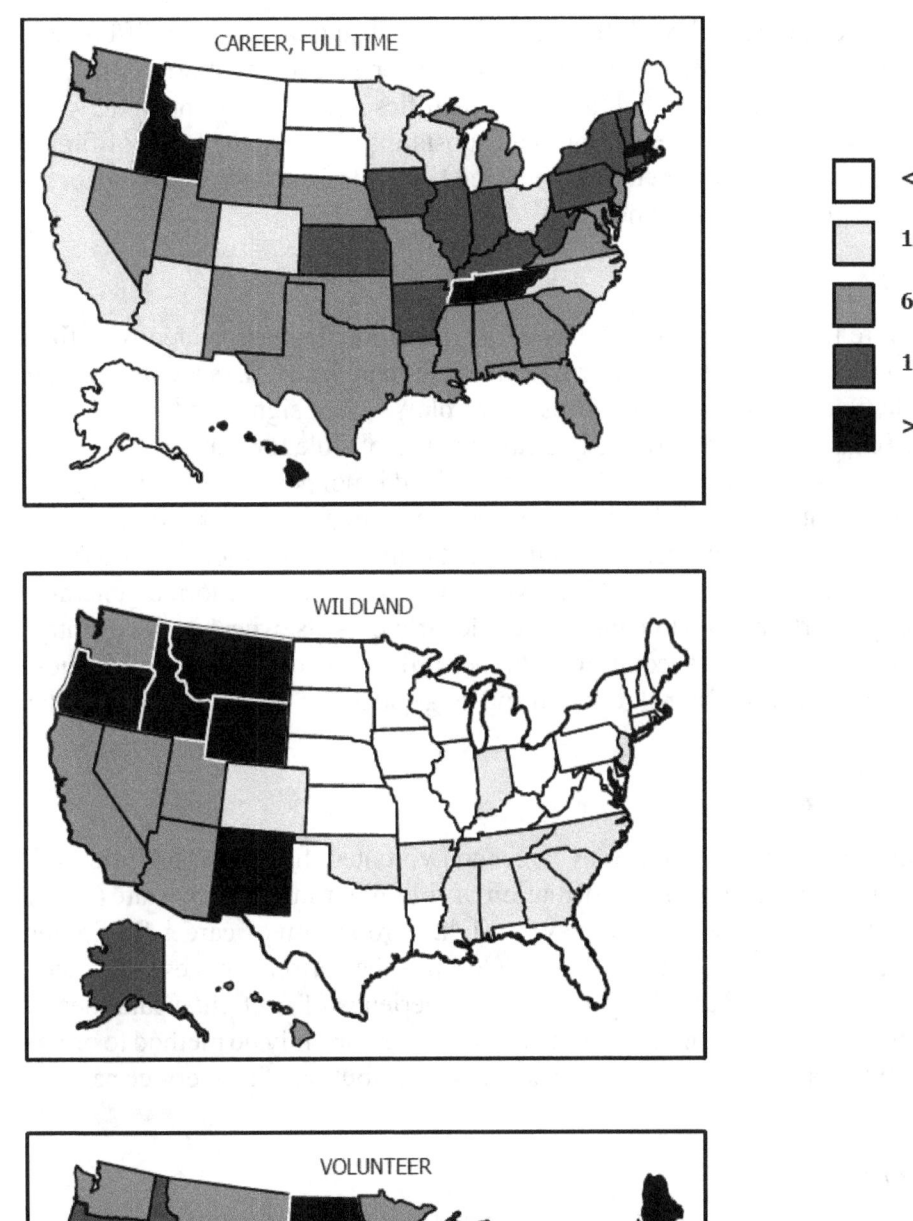

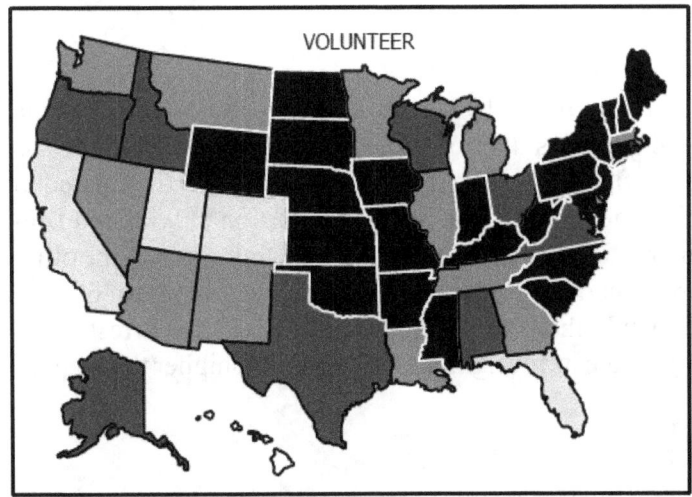

Figure 28. Firefighter Fatalities, by Type of Incident Per 10 Million Population

volunteer, etc.). As an example, although there are clearly substantial climatic differences from one state to another, whether there is a correlation between weather patterns (e.g., precipitation, temperature, humidity, etc.) and firefighter fatalities is not readily apparent. Also, although stricter codes and building regulations can substantially affect the types of fires a state experiences (since these codes are usually enforced at the local level), their impact on the experience of an entire state is not known.

Type of Incident

Although structure fires are generally the leading type of incident associated with firefighter fatalities, this is not the case in California where wildland/brush fires are the leading incident type (Table 6). Similarly, wildland/brush fires play a more significant role in firefighter fatalities in Texas and Florida. Also, EMS calls are responsible for a larger proportion of firefighter fatalities in Florida than elsewhere. Currently, it is not possible to fully explain these differences. Anecdotally, the difference between structure and wildland fires can be explained by climate and fire experience; it is logical that California and Florida experience wildland fires more often than New York. Moreover, fire departments in Florida generally experience a higher proportion of EMS calls than do departments elsewhere in the country, which may explain why such calls account for a higher proportion of firefighter fatalities. These are theories that require additional research and data collection/analysis in order to be proved.

Affiliation of Personnel

As shown in this report, the experiences of career and volunteer firefighter fatalities differ. As expected, there are variations in the affiliation of firefighter fatalities by state (Table 7). Only two states, California and Florida, have a higher proportion of career firefighter fatalities than volunteer. Again, it is not possible to determine the reasons for these variances without additional research. One could argue that the experience of firefighter fatalities is similar to the composition of the state's fire service, but there is currently no method to prove this theory without future analyses that include data on the distribution of fire service personnel by state.

Nature of Fatal Injury

As shown in Table 8, the nature of fatal injury by state also differs by state. California experiences a particularly low rate of fatalities due to cardiac arrest. In part, this may be because only 4 percent of the state's firefighter fatalities are over the age of 60 (compared to about 14 percent of firefighter fatalities generally). The higher proportion of wildland firefighters is also consistent with a higher ratio of deaths due to traumatic injuries. On the other hand, for the MN/WI/MI group, 15 percent of firefighter fatalities are over the age of 60, and that group of states experiences a much higher ratio of deaths due to cardiac arrest. Ultimately, however, California's per capita rate of firefighter fatalities is still higher than that of Michigan, Minnesota, and Florida.

Table 6. Type of Incident by State (1990–2000)

Type of Incident	New York	Pennsylvania	California	Texas	Minnesota, Wisconsin, Michigan	Florida	Ohio
Population	18,483,456	12,081,349	31,815,835	18,919,165	19,391,882	14,460,152	11,100,128
PERCENT*							
Structural Fire/Explosion	56	60	20	37	68	40	47
Wildland/Brush/Grass Fire	2	1	52	25	9	27	10
Nonstructural Fire/Explosion	7	4	0	17	3	0	7
Fire Alarm	4	4	0	0	3	0	0
Motor Vehicle Collisions	20	16	10	10	9	0	10
EMS	3	9	2	4	0	20	10
Water Rescue	0	3	3	0	3	0	3
Confined Space/Trench Rescue	0	0	2	0	0	0	0
False Call	0	1	0	0	0	0	0
Training	7	3	11	8	6	13	13
Total	100	100	100	100	100	100	100

*Column totals may not add to 100 due to rounding.

Table 7. Affiliation by State (1990–2000) (percent*)

Affiliation	New York	Pennsylvania	California	Texas	Minnesota, Wisconsin, Michigan	Florida	Ohio
Paid Full Time	30	22	50	34	27	17	68
Paid Part Time	0	1	0	0	7	11	0
Volunteer	70	77	22	63	66	72	27
Contract	0	0	22	2	0	0	0
Inmate/Prisoner	0	0	3	0	0	0	0
Other (wildland, military	0	0	4	2	0	0	5
Total	100	100	100	100	100	100	100

*Column totals may not add to 100 due to rounding.

Table 8. Nature of Fatal Injury by State (1990–2000) (percent*)

Nature of Fatal Injury	New York	Pennsylvania	California	Texas	Minnesota, Wisconsin, Michigan	Florida	Ohio
Burns/Asphyxiation	19	25	15	25	24	23	14
Cardiac Arrest	57	48	23	36	61	41	47
Internal/Head Trauma	18	23	46	27	10	32	25
Other	6	4	16	12	5	4	14
Total	100	100	100	100	100	100	100

*Column totals may not add to 100 due to rounding.

Comparison of Florida With Pennsylvania

As discussed throughout this report, numerous factors impact the incidence of firefighter fatalities by state. Based on data collected for this study, Table 9 specifically highlights the pattern of firefighter fatalities in Pennsylvania and Florida. Although these states have similar populations (14,460,152 and 12,081,349 respectively), a substantially greater number of firefighters are killed in Pennsylvania than in Florida—97 versus 22 firefighter fatalities during the study period. The hypothesis, therefore, is that since these states have similar populations, factors other than population must affect the incidence of firefighter fatalities.

Table 9. Firefighter Fatality Data Elements: Florida and Pennsylvania

Data Element	Florida		Pennsylvania	
Average Age	39 years		47 years	
Rank	Firefighter Assistant/Deputy Chief Fire Chief	50% 17% 11%	Firefighter Fire Police Officer Assistant/Deputy Chief	67% 16% 5%
Gender	Male Female	95% 5%	Male Female	99% 1%
Nature of Fatal Injury	Cardiac Arrest Trauma (internal or head) Burns/Asphyxiation	41% 32% 23%	Cardiac Arrest Burns/Asphyxiation Trauma (internal or head)	48% 25% 23%

Type of Duty	Extinguishing Fire/
	  Neutralizing Incident
	Other Duties on Emergency Scene
	EMS
		32%
		
		16%
		16%
			Extinguishing Fire/
			  Neutralizing Incident
			En Route/Responding
			Traffic Control
				27%
				
				24%
				13%

Affiliation	Full-Time Career Volunteer Wildland, Full-Time	68% 27% 5%	Volunteer Full-Time Career Paid Part-Time	77% 22% 1%
Type of Agency	Career Volunteer Combination	50% 25% 25%	Volunteer Career Combination	76% 22% 4%
Month of Year	August September July	18% 18% 14%	January August December	18% 10% 10%
Type of Incident	Structure Fire/Explosion Wildland/Grass/Brush Fire EMS	40% 27% 20%	Structural Fire/Explosion MVC EMS	60% 16% 9%
Cause of Fire (if a fire incident)	Arson Natural	60% 40%	Arson Electrical Distribution Heating	40% 15% 10%

Some of the differences between the two states that are reflected in the table include:

- On average, firefighters who die are in Pennsylvania are 8 years older than those in Florida.

- Although cardiac arrest is the leading nature of fatal injury in both states, it affects a higher percentage of firefirghters in Pennsylvania. This may correspond to the older age of Pennsylvania firefighters.

- Significantly more fire police officers die in Pennsylvania than in Florida.

- Significantly more firefighters in Florida are killed while involved in an EMS incident. More Pennsylvania firefighters are killed responding to incidents than Florida firefighters.

- The distribution of career and Volunteer personnel are reversed in the two states. In Pennsylvania, three out of four fatalities are volunteers; in Florida, three of every four fatalities are career firefighters. This difference could be an indication of a factor that affects the total number of firefighter fatalities in Pennsylvania, or it may reflect a larger statewide volunteer firefighter population.

- The leading month for firefighter fatalities in Florida is August, followed by September and July. This is consistent with the prevalence of fatalities involving wildland incidents, which occur predominantly during the summer in Florida. The leading month for firefighter fatalities in Pennsylvania is January, followed by August and December. The higher incidence of winter fires is consistent with the climate in Pennsylvania, where homes must be heated during the colder months. (Heating fires are a leading cause of firefighter fatalities in Pennsylvania.)

Although one could speculate that the composition of the fire service and the distribution of emergency incidents in Florida and Pennsylvania are similar to the distribution of firefighter fatalities, this is not guaranteed. Instead, due to the relatively small sample of firefighter fatalities, one or two anomalous events could significantly alter the data.

To fully address the reasons for the variations discussed in this section, additional research, data, and analyses are required. Some of the causal factors that may be of interest are shown in the adjacent box and may e candidate factors for further research and analyses. (In part, the National Fire Department Census, a project of the USFA, will generate some of these data.) By determining key factors responsible for firefighter fatalities in any state, specific prevention programs could be developed to specifically address and correct identified problems or trends.

Possible Causal Factors for Future Analyses

Climate

Code Enforcement:
Centralized State Office or Local Enforcement
Stringency of Codes Adopted

Average Age of Firefighters

Composition of State Fire Service
Career vs. Volunteer
Gender Distribution

Overall Firefighter Health: Requirement for Annual Physicals?

Training Requirement
Uniform or Different for Career and Volunteer?
Offered at State or Local Level?

Age and Type of Housing Stock

Average Income Level in State

Common Types of Apparatus

Commercial Driver's License Required for Emergency Vehicle Operators?

State/Departmental Participation in NFIRS?
Number of Fire Departments in State
Size of Fire Department

Multiple Firefighter Fatality Incidents

The death of a firefighter has profound effects on the fire service and community as a whole; when more than one firefighter dies in an incident, those effects are magnified. Overall, during the study period, 8 percent of fatal incidents involved the death of more than one firefighter; these incidents accounted for 18 percent of firefighter fatalities. This represents an increase from an earlier USFA study, which found that between 1982 and 1991, only 4 percent of incidents involved the death of more than one firefighter; those incidents accounted for 13 percent of firefighter fatalities [Ref. 9].

Table 10 shows the distribution of firefighter fatalities by the number of firefighters killed per incident. The majority of firefighters are killed in incidents that involve the death of only one firefighter (82 percent). Approximately 14 percent of firefighters are killed in incidents that result in the deaths of two or three firefighters. Incidents involving the death of more than four firefighters are rare, and account for only 3 percent of fatalities.

Table 10. Multiple Firefighter Fatalities

Number of Incidents	Fatalities per Incident	Fatalities	Percent of Firefighter Fatalities
893	1	893	82
59	2	118	11
12	3	36	3
3	4	12	1
2	6	12	1
1	14	14	1
970	30	1,085	100

Fatalities:	1,085
Incidents:	970

Table 11 compares the nature of fatal injuries for firefighters killed in single- vs. multiple-firefighter fatality incidents. Approximately 90 percent of firefighters killed in multiple-fatality incidents die of traumatic injuries (a category that includes internal, head, and other traumas, as well as asphyxiation, burns, and burns/asphyxiation). In contrast, only 37 percent

Table 11. Nature of Fatal Injury, Single- vs. Multiple-Fatality Incidents

Nature of Fatal Injury	Firefighters in Multiple-Fatality Incidents	Firefighters in Single-Fatality Incidents
Asphyxiation	32%	7%
Burns	12%	2%
Burns/Asphyxiation	15%	2%
Cardiac Arrest/ Heart Attack	1%	53%
Drowning	3%	2%
Electric Shock	3%	1%
Gunshot	3%	0%
Trauma	32%	26%
Stroke/CVA	0%	2%
Total	100%	100%

Multiple Fatality		Single Fatality	
Sample size:	190	Sample size:	895
Valid entries:	188	Valid entries:	886
Invalid entries:	2	Invalid entries:	9

of those killed in single-fatality incidents die of traumatic injuries; rather, they are more likely to die from heart attacks.

Self-Contained Breathing Apparatus Depletion

Self-contained breathing apparatus (SCBA) are designed to provide firefighters with air while they are operating in dangerous environments. How quickly firefighters deplete an SCBA's air supply depends on the size of the cylinder used, the effort they exert, and their level of physical fitness. During extremely physically demanding activities, a firefighter could deplete an SCBA in as little as 10–15 minutes (when using a 30-minute air cylinder).

In 16 percent of fatalities during the study period (178 cases) firefighters were reportedly wearing an SCBA at the time of their injury; 30 percent had reportedly depleted their SCBA air supply. For the remaining 70 percent of fatalities, the air supply was either not depleted or that information was unavailable.

Personal Alert Safety System Device Activation

Personal Alert Safety System (PASS) devices were originally developed in the early 1980s. They are usually clipped to the harness of a firefighter's SCBA and are designed to emit a loud signal if a firefighter becomes trapped or incapacitated while operating on a fire-ground. A pre-alert signal goes off 10 seconds before the alert signal and can be disabled by the firefighter's moving; if the device has not sensed motion in 30 seconds or the device is manually activated, the alert signal is emitted.

In the 89 cases during the study period where a firefighter was reportedly wearing a PASS device at the time of their fatal injury, the device activated only 9 percent of the time. In 44 percent of incidents the device reportedly was worn but did not activate; in the remainder of incidents, it is unknown whether or not the device activated.

The effects of PASS devices on firefighter mortality are not yet clear. In its investigation of incidents involving firefighter fatalities, the National Institute for Occupational Safety and Health (NIOSH) has found multiple occasions where firefighters either failed to wear or activate their PASS devices prior to entering a structure [Ref. 10]. To combat this trend, SCBA manufacturers are incorporating PASS devices into their SCBA systems. Such devices are turned on automatically when the air cylinder is opened. As with standalone PASS devices, integrated PASS devices will alert if it does not sense motion. There is no way to disable the device while the SCBA system is pressurized. (There is a reset button in the event that the device activates accidentally.) It will be interesting to see how these new technologies affect firefighter mortality in the future. Revisiting this topic in 5–10 years could show significant change in the use and activation of PASS devices.

Wildland Firefighters

Wildland firefighting are often prominently featured in the media due to several high-profile incidents involving wildland firefighter fatalities and the intense wildfire season in 2000. Table 12 compares the nature of fatal injury for wildland firefighters vs. non-wildland firefighters. Wildland firefighters are far more likely to be killed by traumatic injuries than are non-wildland firefighters. Conversely, they are significantly less likely to die of a heart attack for a variety of reasons including that wildland firefighting agencies typically have extremely high standards of physical fitness. Wildland firefighters also tend to be younger than non-wildland firefighters; nearly 70 percent of part-time wildland firefighters are under the age of 30.

For further information on wildland firefighter safety, contact the U.S. Forest Service or see *http://www.fs.fed.us/fire/infodir.shtml*; and *http://www.usfa.fema.gov/pdf/ wildff90–98.pdf*.

Table 12. Nature of Fatal Injury, Wildland Firefighters vs. Non-Wildland Firefighters (1990–2000)

Nature of Fatal Injury	Wildland*	Non-Wildland
Asphyxiation	2%	12%
Burns	4%	4%
Burns/Asphyxiation	23%	3%
Cardiac Arrest/Heart Attack	7%	47%
Drowning	1%	2%
Electric Shock	2%	2%
Trauma	58%	24%
Other Medical	1%	1%
Stroke/CVA	1%	1%
Total	100%	100%

*Including contract personnel and prisoners

Wildland Fatality	
Valid entries:	83
Valid entries:	83
Invalid entries:	0

Non-Wildland Fatality	
Sample size:	1,002
Valid entries:	992
Invalid entries:	10

PREVENTION OF FIREFIGHTER FATALITIES

Some forces and circumstances that lead to firefighter fatalities are simply beyond human control. However, through research, study, training, improved operations, development of new technologies, the appropriate use of staffing, and other factors, it should be possible to significantly reduce the number of firefighters killed each year. Moreover, firefighter fatalities are generally the result of a chain of events, which, if detected early, may be broken to prevent many, or even most, fatalities.

This chapter addresses prevention strategies for the fire service. Given their persistent nature, four specific topics are discussed in detail: pre-existing conditions, training, fireground, and motor vehicle collisions. Further information on the prevention of firefighter fatalities can be found on the USFA website at http://www.usfa.fema.gov.

To enhance the overall safety of all members of the fire service, personnel should take care to adhere to standard operating procedures and maintain their personal health and fitness. This includes maintaining proper hydration on the fireground and wearing seatbelts at all times while operating or riding apparatus (or POVs). Moreover, at a higher level in the chain of events, strong public education programs can prevent emergency incidents from happening in the first place; were the incident prevented, the risks would be mitigated.

Pre-Existing Conditions

Previous existing medical conditions that affect the health and safety of firefighters include underlying medical diseases and the state of physical fitness of the firefighter at the time of his or her death. These deaths are different from other on-duty deaths in that the injury may not have occurred or been fatal to that individual under the same conditions in the absence of the pre-existing condition.

Firefighter Physical Fitness—In August 2000, the NFPA released NFPA 1583: *Standard on Health-Related Fitness Programs for Fire Fighters* [Ref. 11]. It establishes the minimum requirements for a health-related fitness program for fire department members who are involved in rescue, fire suppression, EMS, hazardous materials operations, special operations, and related activities. The standard requires fire departments to establish a "health-related fitness program that enables members to develop and maintain a level of health and fitness to safely perform their assigned functions." Departments must appoint a health and fitness coordinator (HFC) to coordinate the program and perform periodic fitness assessments to determine an individual's exercise needs.

A 2001 study at the Applied Exercise Science Laboratory at Texas A&M University investigated firefighters' risk of suffering a heart attack. The study followed 74 firefighters ages 20–60 years old over a 6-year period and concluded that firefighters have long periods of stress-free activity during the day, and when the call for help comes, there is a "sudden, intense energy demand required, and if they are not in adequate physical condition, the results can be deadly" [Ref. 12].

A 1996 study performed by the Montgomery (AL) Fire Department in cooperation with The Human Performance Laboratory at Auburn University found that added weight and body fat affected performance of firefighters on the fireground. The research showed that there was a direct relationship between added body weight and decreased physical performance. Additionally, as body weight increased, efficiency decreased and fatigue set in faster. Today, the Montgomery Fire Department monitors firefighters' height, weight, and body fat composition at various points throughout their careers [Ref. 13].

Many fire departments are implementing mandatory physical fitness programs. The Oklahoma City Fire Department implemented a mandatory, on-the-job exercise program. These firefighters are required to exercise for 1½ hours on each 24-hour shift. Officials with the Oklahoma City Fire Department hope that by implementing the workout program, firefighters will live longer and avoid heart attacks. Firefighters meet with a physician at the beginning of the program and are then assigned to a physical wellness coordinator. The coordinator runs through a series of exercises to see what shape the firefighter is in and then, based on the results of the tests, creates an aerobic and weightlifting regimen. Firefighters are reevaluated every 6 months. Officials with the department have noted that the overall fitness of the department has improved with the implementation of the program [Ref. 14].

A variety of commercial fitness programs are widely available. Such programs include strength building, endurance, and flexibility exercises as well as tips on nutritional habits. Programs like step aerobics and Tae-Bo are becoming popular in fire departments across the country. These programs incorporate fun into cardiovascular exercise, motivating participants to continue with the workout regimen. Where possible, fire departments should incorporate physical training into the firefighter's daily schedule. This should be done in conjunction with fitness experts, physicians, and nutritionists to ensure that firefighters get in shape in a safe, healthy manner.

The IAFF, International Association of Fire Chiefs (ICHIEFS), and 10 pairs of local unions and their municipalities joined together to form the Fire Service Joint Labor Management Wellness–Fitness Task Force. The task force created the Fire Service Joint Labor Management Wellness–Fitness Initiative in an attempt to build a stronger and healthier fire service. The initiative is a fitness program that includes physical, physiological, and psychological components. The program comes complete with a physical fitness and wellness program package that includes a manual and a video. It is hoped that all departments affiliated with the IAFF will implement the program [Ref. 15].

While it might appear that most physical fitness programs are targeted to career departments, many volunteer departments are encouraging their members to be more health conscious. To facilitate this, some departments have built weight rooms with state-of-the-art fitness equipment for use by their members. For departments that cannot afford expensive equipment, other options include soliciting donations of used exercise equipment. Other departments have approached commercial gyms or local recreation centers and formed agreements granting members free or reduced-price memberships. Some departments have hired fitness consultants who meet with members on a regular basis to develop individual and department-wide fitness programs.

Candidate Physical Agility Test—One way to prevent firefighters from having poor fitness habits is to recruit and hire firefighters with good fitness habits. Hiring healthy individuals to serve as firefighters may reduce firefighter fatalities from heart attacks and other medical conditions; physically fit individuals may also be at less of a risk of incurring traumatic injuries. However, a fair standard must be applied to all applicants.

The members of the Fire Service Joint Labor Management Wellness–Fitness Task Force developed the candidate physical agility test (CPAT) to establish a nondiscriminating, fitness-based test for hiring firefighters. The CPAT is administered along with other recruiting and mentoring practices. The CPAT was designed for the recruitment process for career departments, but it can also be applied during recruitment of volunteer firefighters.

When practical, departments should have EMS personnel and equipment standing by during the actual test should any candidate suffer a medical problem or injury.

The CPAT is comprised of eight events in which the candidate must wear a 50-pound weighted belt. (A belt is used as opposed to structural turnout gear and SCBA so as not to give an advantage to experienced firefighters seeking employment.) The eight events include:

- Stair climb—climbing stairs with a 25-pound simulated hose pack
- Ladder raise and extension—placing and raising a ground ladder to the desired floor or window
- Hose drag—stretching and advancing hoselines, charged and uncharged
- Equipment carry—removing and carrying equipment from fire apparatus to fire-ground
- Forcible entry—penetrating a locked door, breaching a wall
- Search—crawling through dark areas to search for victims
- Rescue drag—victim removal from a fire building
- Ceiling pull—pulling a ceiling to check for and locate fire extension.

By incorporating the CPAT program into the recruiting and hiring process, fire departments increase the chance they will build a membership that is physically fit and able to fulfill the demanding duties of firefighters.

For further information on the efforts of the Fire Service Joint Labor Management Wellness–Fitness Task Force, contact the IAFF (http://www.iaff.org) or IAFC (http://www.iafc.org).

Heart Attacks—Heart attacks are the leading cause of firefighter fatalities. The physical demands placed on firefighters can be very high and they often have to go from a state of deep sleep to near 100 percent alertness and high physical exertion in a matter of minutes. Further, they must carry heavy equipment through intense heat while wearing heavy gear. Due to the physical demands of firefighting, firefighters must maintain a high level of physical fitness.

The medical term for heart attack is myocardial infarction [Ref. 16]. A heart attack occurs when the blood supply to part of the heart muscle itself—the myocardium—is severely reduced or stopped. This occurs when one of the coronary arteries that supply blood to the heart muscle is blocked. The blockage is often from the buildup of plaque (deposits of fat-like substances) known as atherosclerosis, combined with arteriosclerosis (the progressive hardening and thickening of the arteries). During exercise or excitement, the narrowed coronary arteries caused by arteriosclerosis cannot increase the blood supply enough to meet the increased oxygen demand of the heart muscle. The plaque can eventually tear or rupture, triggering a blood clot to form that blocks the artery and leads to a heart attack. If the blood supply is cut off severely or for a long time, muscle cells suffer irreversible injury and die. Disability or death can result, depending on how much of the heart muscle is damaged.

At autopsy, a substantial number of firefighter fatalities were found to have severe coronary arteriosclerosis. Family members reported that many had been diagnosed with hypertension or diabetes prior to their deaths. Many factors affect a person's risk of suffering a heart attack. Some risk factors are outside of an individual's control such as increasing age, gender (male), and heredity (including race). Other risk factors, however, can be controlled or modified through diet, exercise, and personal choice. Modifiable risk factors include use of tobacco, high cholesterol, high blood pressure (hypertension), physical inactivity, obesity, and diabetes. The following are examples of ways fire departments can encourage their members to improve their health and lower their risk.

In a significant number of cases, no signs or symptoms precede heart attacks. In such cases, patients fall victim to sudden cardiac death (also called sudden death). Sudden cardiac death is the result of an abrupt loss of heart function. Victims of sudden cardiac death may or may not have a previous medical history of heart disease. All heart diseases can lead to cardiac arrest or sudden cardiac death. Many cardiac arrests occur when electrical impulses in the heart become rapid (ventricular tachycardia) or chaotic (ventricular fibrillation). These cardiac arrhythmias cause the heart to stop. Cardiac arrest can lead to brain death, and ultimately clinical death, if not treated within 4 to 6 minutes. The definitive treatment for these lethal arrhythmias is rapid defibrillation.

The American Heart Association website is an excellent resource for educational materials, prevention strategies, and other information on heart attack and stroke (http://www.americanheart.org). *Firehouse Magazine's* website (http://www.firehouse.com/fitness/) has a page dedicated to fitness and well being. The page offers a variety of workouts aimed to help firefighters with such things as strength training, flexibility training, and cardiovascular conditioning, and it provides information on the latest trends in health and fitness.

Diet—Meals that are high in fat, cholesterol, sodium, and calories can contribute to or exacerbate health problems such as heart disease and diabetes. Maintaining a healthy diet can mitigate these health issues and reduce firefighters' risk.

Smoking—Smokers have twice the risk of dying from heart disease than nonsmokers [Ref. 17]. Smoking reduces lung capacity and narrows blood vessels, damaging lungs and reducing the amount of oxygen available during strenuous activities, including firefighting. Smoking also causes chronic and potentially fatal lung diseases such as bronchitis and emphysema.

Smoking-related diseases and heart attacks are preventable. In the years after a smoker quits, their risk of dying from a heart attack decreases markedly. Fifteen years after quitting, an ex-smoker faces the same risks as someone who has never smoked [Ref. 17]. Also, their risk of lung disease, heart disease, stroke, diabetes, and hypertension all decrease. Although these lifestyle changes directly affect the firefighter's individual health, they also benefit the department as a whole, as healthier firefighters may be less likely to be injured or killed while on duty.

The most effective way to prevent smoking-related disease is to maintain a smoke-free environment, which can be accomplished in several ways. Some fire departments have banned personnel from smoking while on duty; others have attempted to ban firefighters from smoking both on and off duty.

As an alternative, some departments have offered to pay for smoking cessation programs to help their firefighters quit smoking. These programs include the use of nicotine gums, patches, medications, various forms of psychotherapy, or a combination of these approaches. Offering a smoking cessation program is usually less objectionable to firefighters than ordering them to quit. Promoting a smoking cessation program can be cost effective to a fire department. By promoting such a program, fire departments make an investment in their future that may cost a down payment for the program itself, but should theoretically lower the cost of future medical care for that firefighter and for the department.

Alcohol—In a recent study by the Centers for Disease Control and Prevention, nearly 52 percent of Americans over the age of 12 had consumed an alcoholic beverage in the month prior to the survey [Ref. 18]. Many people, including firefighters, occasionally consume alcoholic beverages. However, the adverse effects of excessive alcohol consumption can be deadly.

Alcohol also affects the heart and, in high amounts, can produce irregular heartbeats. Over time, a rapid or irregular heartbeat can lead to hypertension, heart failure, stroke, or other complications [Ref. 16]. Excessive alcohol consumption also causes pancreatitis, nutritional deficiencies, malignancies, and cirrhosis.

The fire service must be aware of the signs of alcohol abuse and make resources available to members with alcohol problems. For further information, contact the National Institute on Alcohol Abuse and Alcoholism (http://www.niaaa.nih.gov) or Alcoholic's Anonymous (http://www.alcoholics-anonymous.org).

Employee Assistance Programs (EAPs)—Both career and volunteer firefighters can benefit from the assistance of EAPs, which are programs organized to provide counseling, support, and other services. Some larger fire departments have their own EAPs; smaller agencies (of all types) might consider contracting for EAP services or participating in a local government's EAP.

Training

Seven percent of firefighters deaths occur during training exercises each year. To prevent these fatalities, departments must approach training with safety as the leading priority.

Physical Fitness—The majority of deaths during training occur while the firefighter is engaged in physical fitness activities. The paradox in this situation is that firefighters must engage in physical activities to perform their duties effectively. To prevent deaths related to physical fitness, firefighters should engage in such activities only after being evaluated by a physician. With the help of the medical community plus certified physical trainers, lower impact exercise programs can be developed that improve the firefighter's level of fitness but pose a lower risk of causing injury or death.

Training Exercises—Training exercises should include didactic as well as classroom sessions to prepare firefighters for the scenarios to be used in the practical exercise. Practical training sessions should be conducted in accordance with departmental SOPs; applicable federal, state, and local laws; and industry standards. Only those personnel who are qualified should be allowed to directly participate in training activities. Personnel that lack the training or qualifications to participate should be encouraged to assume support functions. Allowing unqualified personnel to participate in training or actual incidents could expose a fire department to serious liability in the event of an injury or fatality.

Since the purpose of practical training is to familiarize personnel with the conditions they will face in an actual incident, full protective equipment, including SCBA, PASS devices, and accountability systems should be used where appropriate. In particular, live-fire evolutions should conform to NFPA 1403, *Standard on Live Fire Training Evolutions in Structures* [Ref. 19].

Fireground

Fireground firefighter fatalities occur on the scenes of actual emergencies and tend to occur as the result of flashovers, structural collapses, and falls. One of the highest priorities a department should strive for and reinforce is familiarity with basic firefighting skills and self-rescue techniques. Familiarity with these techniques can reduce the likelihood that firefighters will become lost in a structure and require rescue.

Incident Command System (ICS)—The ICS is the model tool for the command, control, and coordination of resources and personnel at the scene of emergencies. It was designed to facilitate operations and efficient mitigation at the scene of an emergency, regardless of its magnitude, location, duration, or nature. The ICS is designed to allow for an incident commander (IC) to be identified regardless of that individual's rank. The system allows for the progressive transfer of command from lower ranking individuals to higher ranking individuals as necessary.

Under the ICS, the IC designates sector commanders to assume responsibility for particular aspects of an emergency response. The IC determines the overall strategic objective for the incident; the sector commanders are responsible for developing tactical plans of action to achieve that strategic goal. Examples of sectors include command, EMS, triage, rescue, logistics, transportation, finance, and safety. For the safety sector, at a minimum, the IC should designate a safety officer who has the authority to alter, suspend, or terminate any activity at the scene if he or she determines those actions to be imminently unsafe [Ref. 20].

Freelancing occurs when firefighters or companies deviate from their assigned duties. What makes freelancing so dangerous is that the IC may give orders to another company that endangers the freelancers, or the freelancers may operate in a manner that endangers other firefighters. Training in ICS and operations may mitigate the occurrence of freelancing.

Many fire departments currently implement an ICS for working incidents. This practice should continue; those departments that do not currently use an ICS should consider developing one that suits their needs.

Accountability—As part of his or her responsibilities, the IC needs to know the approximate location of companies and personnel operating at the incident. While sector commanders accomplish part of this through regular reports to the IC, a structured Personnel Accountability System is also a necessity. Some Personnel Accountability Systems use technology, such as barcoding, to track the assignments of personnel and apparatus. Other Personnel Accountability Systems are simpler, and use plastic tags or dry-erase boards to manually track activities on the scene.

Closely related to Personnel Accountability Systems are PASS devices. As discussed previously, these devices are designed to emit a loud tone in the event that a firefighter becomes incapacitated or otherwise unable to move while operating at an incident. When activated, these devices can help search and rescue teams locate a downed firefighter quickly. In particular, if the team knows where the firefighter or company was assigned to operate, the sound of an activated PASS may lead them to the downed firefighter(s). NFPA 1982: *Standard on Personal Alert Safety Systems* [Ref. 21] specifies the requirements for PASS devices. Fire departments should make sure that PASS devices used comply with NFPA 1982.

In conjunction with a formal Personnel Accountability System, many departments also incorporate Personnel Accountability Reports (PARs) into their SOPs for emergency operations. PARs are generally conducted over the radio and require each company or firefighter to verbally check in with the IC at regularly scheduled intervals. Often, the dispatch center will advise the IC at certain points during the incident (e.g., 15 minutes, 30 minutes, 1 hour) to perform an accountability check. Once it is determined that a firefighter or company is not accounted for, the IC can shift resources from suppression or mitigation to search and rescue.

It is crucial that personnel are familiar with the particular Personnel Accountability System and PASS devices used by their departments. This requires ongoing training and cooperation across jurisdictions where mutual aid response is common.

Tactical Strategies—More fire departments are adjusting their tactics to allow for defensive operations when no life hazards are present. While some have criticized this shift, it may prove crucial in reducing the number of future firefighter fatalities. This does not mean that all aggressive interior firefighting should stop. Rather, ICs should be aware that in instances where the layout of a structure cannot be ascertained with certainty or there are hazardous substances present, an exterior attack should be considered, unless there are confirmed victims trapped in the structure. Firefighters have been killed searching for victims who have already exited a building or were never there in the first place.

Two-In/Two-Out—According to the Occupational Safety and Health Administration (OSHA) policy (29 CFR 1910–134) for interior structural firefighting, firefighters must be in direct contact with at least one other firefighter. Direct contact includes being able to maintain verbal or visual contact with other firefighters inside an area immediately dangerous to life and health (IDLH) at all times. Additionally, OSHA requires that a team of at least two properly equipped and trained firefighters be present outside of the IDLH area before any teams enter the structure. For further information, see http://www.osha-slc.gov/OshStd_ data/1910_0134.html.

Related to Two-In/Two-Out is the development of rapid intervention teams/crews (RITs/ RICs). NFPA 1710: *Standard for the Organization and Deployment of Fire Suppression Operations, Emergency Medical Operations, and Special Operations to the Public by*

Career Fire Departments and NFPA 1720: *Standard on Volunteer Fire Service Deployment* recommend that at least two trained firefighters be assigned to an RIT. Although initial RITs often consist of only two firefighters (to comply with the two-out of Two-In/Two-Out), firefighter rescue requires many trained hands, especially if multiple firefighters are in distress at the same time.

RITs should stand by on the fireground and be deployed immediately when interior firefighters find themselves in imminent danger (e.g., they are unable to escape and continued presence in the current environment may lead to serious injury or death). Such teams should be equipped, at a minimum, with an extra SCBA, rope, forcible entry tools, a hoseline (from an engine not supplying the primary attack lines), and, if possible, a thermal imaging camera. As soon as a "Mayday" is transmitted, command should acknowledge the transmission and immediately deploy the RIT. Other units operating on the scene should be switched to an alternate tactical channel to allow for direct communication with the missing firefighter(s). For further information on the development of RITs, contact the USFA.

Self-Rescue—Firefighters who find themselves in imminent danger and are unable to wait for a RIT rescue may be able to escape from a structure using some crude but effective escape techniques. These techniques must be practiced, like any other techniques, but since they are often performed from an elevated level, fall protection and qualified instructors must be employed. Some examples of personal escape taught by various agencies and departments in the fire service include charged hose line slide, rappel with a personal escape rope, and wall breach. These are only some techniques being taught, not all of which are universally accepted as "best practices."

Salvage and Overhaul—Salvage and overhaul operations begin during the extinguishment of a fire and continue after the fire has been knocked down. Though there might not be any fire in the building during salvage and overhaul operations, the atmosphere remains hazardous.

The structure in areas directly above, adjacent to, and in the room of fire origin, may have been seriously compromised by fire spread, smoke, and water. Firefighters and the safety officer should be alert to recognize when the integrity of the structure is compromised. Areas being salvaged often are still smoky and usually are filled with toxic fire byproducts of or gases not ignited during the fire. According to a study published by researchers at the University of Arizona College of Public Health, firefighters not wearing respiratory protection during salvage and overhaul were significantly more prone to lung injury than those that did wear such protection [Ref. 22]. Therefore, firefighters must be in full personal protective equipment (PPE), including SCBA, at all times while performing salvage and overhaul.

Motor Vehicle Collisions

Firefighter fatalities as a result of MVCs are an ongoing problem and accounted for 22 percent of the total number of firefighter fatalities from 1990–2000.

Driver Training—Driver familiarity with fire apparatus facilitates comfort and control over the vehicle. But driver familiarity must also be supplemented with a formal driver training program that educates the driver about obstacles, hazards, and roadway types.

In addition to licensing requirements at the state level, many fire departments require drivers to successfully complete an emergency vehicles operator's course (EVOC) and departmental checkout programs prior to driving apparatus. EVOC involves both didactic and practical exercises to teach firefighters how to drive in emergency response situations. The course addresses topics such as high-speed driving, evasive maneuvering, and skid control.

Departmental checkout programs familiarize drivers with the particular vehicles they will be operating on a regular basis. This program should involve driving on the different types of roadways in the department's first-due area. This might include in-town driving, out-of-town (rural) driving, highway driving, and off-road driving. If a department protects an area with dense traffic, the driver should train on maneuvering through, in, and around heavy traffic. The use of seatbelts should be stressed.

Driver training programs can be measured by the duration of time spent performing driver training, distance driven during driver training, or a combination of both. Throughout the program, the driver trainer must be unbiased in his/her evaluation of the driver trainee. Driver training, like all skills, should be practiced on a regular basis. In particular, tanker drivers should practice maneuvering the apparatus with the tank at various fullness levels. It is crucial that operators train regularly on the operation of tankers to prevent fatal errors while responding to incidents.

Fire departments should consult NFPA 1002: *Fire Department Vehicle Driver/Operator Professional Qualifications* [Ref. 23] when developing or modifying its driver training program. This standard addresses the knowledge and skills needed to operate and maintain fire department vehicles.

Dispatch and Response Criteria—Fire departments establish dispatch and response criteria based on applicable statutes, local precedent, and available resources. Responding with multiple apparatus to an incident may bring a lot of equipment to the scene, but not all of it may be necessary. Response guidelines should be established according to the nature and type of incident and the personnel needed to mitigate that incident. In some instances, a single-unit response may be appropriate rather than sending an entire first-alarm assignment.

Additionally, fire departments must consider whether an emergency response (using lights and sirens) to every incident is warranted. Many departments have shifted to emergency responses only for life-threatening or potentially life-threatening emergency incidents and not for routine responses for non-life-threatening incidents. Emergency vehicles traveling with emergency warning lights and sirens activated have a significantly higher risk of being involved in a motor vehicle collision [Ref. 24].

Intersections—The majority of MVCs occur at intersections; drivers should approach intersections with extreme caution, coming to a complete stop before entering the intersection [Ref. 25]. The driver should proceed through the intersection only when he/she is positive that the intersection is clear and that all opposing traffic has come to a complete stop. Some departments have prohibited their fire apparatus from entering intersections on a red light. These departments require their drivers to wait until a green light appears before proceeding through it.

Some departments are using modern technology to minimize the delay caused by red lights. These departments are installing equipment that provides the right-of-way for emergency vehicles at signalized intersections. Using high-energy infrared technology, the mounted device remotely toggles traffic lights from red to green. Departments that use these require their drivers to wait until their light turns green before proceeding. If the light does not turn green within a specified period of time, it means that another fire apparatus is approaching from a different direction. Such technology facilitates maneuvering through intersections more safely, helping to reduce the number of collisions involving fire apparatus.

Seatbelts—As noted earlier, the majority of firefighters killed in MVCs were not wearing seatbelts. Numerous studies have proven that wearing lap and shoulder belts reduce the risk of fatal or serious injury in a collision. When responding to emergency calls, drivers must do so with safety as their number one priority. The apparatus driver and unit officer should make sure that everyone aboard the apparatus is wearing their seatbelt before moving the vehicle. According to the National Highway Traffic Safety Administration, 49 states, the District of Columbia, and a number of American territories have enacted seatbelt laws. Although some states have exemptions for public safety agencies, the fact remains that seatbelts save lives, whether one is riding in a POV or a piece of fire apparatus.

Privately Owned Vehicles—Some departments operate on an on-call system where members respond to the fire station or fire scene in their POV. More firefighters are killed each year while responding to emergency calls in their POVs than any other vehicle. In some cases, personnel may break traffic laws like exceeding posted speed limits or rolling through stop signs while trying to get to an emergency. Not all states allow emergency lights on POVs; rather, some allow firefighters to use "courtesy lights." Other vehicles are not legally required to yield the right-of-way to vehicles with courtesy lights and vehicles outfitted with such equipment are required to obey all traffic laws.

Firefighters should check with the appropriate agencies to determine what types of lights, if any, are permitted in their state. Where emergency or courtesy lights are permitted, firefighters must be vigilant about operating their POVs in a safe, responsible manner, regardless of the nature of the emergency to which they are responding.

RESOURCES

The deaths of firefighters have far reaching effects. It is very important that fire departments have a written plan in the event of the on-duty death of a firefighter. The NFFF has many resources to help departments, chief officers, and surviving family and friends. These include stress counseling, participation in the National Fallen Firefighters' Memorial Weekend, and financial planning. Fire service members should also take stock of their personal finances to ensure that their families will be adequately supported in the event of their untimely death. Moreover, they should draft a will that includes their wishes for funeral/memorial services (e.g., a service funeral with apparatus or a private, family memorial).

Many local churches, religious organizations, and charities (including the Red Cross) make themselves available to families and fire departments in the aftermath of a fatality. The IAFF and its local chapters sponsor various programs for the families of deceased firefighters and often coordinate the memorial service/funeral for the deceased. Volunteer firefighters' associations coordinate similar programs for their members.

Many larger fire departments have their own counselors on staff to assist families after the death of a firefighter. Departments without these resources may consider using the assistance of local chaplains or representatives of the local police department. Mental health clinics, hospitals, and private mental health professionals are also valuable resources for helping families and firefighters cope with their loss.

This chapter addresses some specific resources available to families and fire departments. For further information, contact the NFFF at http://www.firehero.org.

Fire Service Resources

General Resources

Taking Care of Our Own: A Resource Guide is a 50-page guide contains information on pre-incident planning, notification, family and fire department support, and resources for fire departments. The publication is available free of charge.

> The National Fallen Firefighter Foundation
> P.O. Drawer 498
> Emmitsburg, MD 21727
> (301) 447–1365
> http://www.firehero.org/fire/comp_orderform2.htm

The Aftermath of Firefighter Fatality Incidents: Preparing for the Worst. United States Fire Administration, Technical Report Series, Report 089.

> United States Fire Administration
> Publications Center
> 16825 South Seton Avenue
> Emmitsburg, MD 21727
> 1–800–561–3356
> http://www.usfa.fema.gov

Firefighter Fatality Reports

Firefighter Fatalities in the United States. FEMA, USFA. Reports can be downloaded for the years 1986–2000 at the Website address.

> United States Fire Administration
> Publications Center
> 16825 South Seton Avenue
> Emmitsburg, MD 21727
> 1–800–561–3356
> http://www.usfa.fema.gov/nfdc/ff_fat.htm

Wildland Firefighter Fatalities in the United States 1990–1998. United States Department of Agriculture, Forestry Service. The report can be downloaded at the Website.

> http://www.usfa.fema.gov/pdf/wildff90-98.pdf

NIOSH fire fighter fatality programs and reports, firefighter fatality investigation and prevention program, and other reports are available for many incidents and can be downloaded from the Website.

> National Institute for Occupational Safety and Health
> 4676 Columbia Parkway
> Cincinnati, OH 45226
> 1–800–35–NIOSH
> Fax: (513) 841–4488
> http://www.cdc.gov/niosh/facerpts.html

Investigations

Firefighter Autopsy Protocol, United States Fire Administration, 1991. Recommended procedures on investigating the cause and contributing factors of a firefighter's death for coroners, medical examiners, and pathologists.

United States Fire Administration at http://www.usfa.fema.gov or

United States Fire Administration
Publications Center
16825 South Seton Avenue
Emmitsburg, MD 21727
1–800–561–3356
http://www.usfa.fema.gov/pdf/usfapubs/fa-156.pdf

Guide for Investigation of a Line of Duty Death, IAFC. This guide, developed by ICHIEFS Health and Safety Committee, contains procedures and strategies for detailed investigations of line-of-duty deaths and injuries.

International Association of Fire Chiefs
4025 Fair Ridge Drive
Fairfax, VA 22033–2868
(703) 273–0911
publications@iafc.org;
http://www.ichiefs.org/departments/pubs.htm#lod

Standard Operating Procedures for On-Duty Deaths

The NFFF provides SOPs and guidelines that were developed by fire departments for use in the event of the loss of a firefighter.

The National Fallen Firefighter Foundation
P.O. Drawer 498
Emmitsburg, MD 21727
(301) 447–1365
http://www.firehero.org/

A Procedural Guide in the Event of Death in the Line of Duty of a Member of the Volunteer Fire Service, 1987. This 30-page book covers proper procedures to follow in the event of a on-duty death or injury.

> National Volunteer Fire Council
> 1050 17th Street, NW
> Suite 490
> Washington, DC 20036
> (202) 887–5700
> 1–888–ASK–NVFC
> Fax: (202) 887–5291
> nvfcoffice@nvfc.org
> http://www.nvfc.org/manuals.html

Stress and Grief

International Critical Incident Stress Foundation, Inc., is a nonprofit organization that specializes in prevention and treatment of disabling stress for emergency services.

> International Critical Incident Stress Foundation, Inc.
> 10176 Baltimore National Pike
> Unit 201
> Ellicott City, MD 21042
> 24-Hour Emergency Hotline: (410) 313–2473
> Phone: (410) 750–9600
> Fax: (410) 750–9601
> http://www.icisf.org/

Financial Assistance for Children and Spouses

The following are examples of programs offered at the national level. For further information about resources available, by state, contact the NFFF.

Nationwide

The PSOB's program provides financial assistance for firefighters who are totally disabled or killed by a traumatic injury in the line of duty.

> Public Safety Officers' Benefits Program
> 810 Seventh Street, NW
> Washington, DC 20531
> (202) 307–0635
> 1–888–SIGNL13 (744–6513)
> Fax: (202) 307–3373
> http://www.ncjrs.org/pdffiles1/fs000066.pdf

Private

The Firefighters Safety Association provides a benefit to the families member firefighters who are killed in the line of duty. In addition, the Association provides assistance in applying for the PSOB program, reporting the death to the National Fire Academy, and offers other forms of support.

Firefighters Safety Association, Inc.
33 Page Avenue, Suite 201
Asheville, NC 28801
(866) 253–9546
info@FirefighterSafety.net
http://www.firefightersafety.net/benefits.html

The National Volunteer Fire Council provides an accidental death and disability life insurance policy to its members.

National Volunteer Fire Council
1050 17th Street, NW
Suite 490
Washington, DC 20036
(202) 887–5700
1–888–ASK–NVFC
Fax: (202) 887–5291
www.nvfc.org/manuals.htm
nvfcoffice@nvfc.org

VFIS is an insurance company that specializes in providing insurance services to volunteer firefighters as well as other emergency service organizations. VFIS provides a wide range of services, including accident and sickness, disability, death, and retirement benefits.

VFIS Main Offices
183 Leader Heights Road
P.O. Box 2726
York, PA 17405
1–800–233–1957
(717) 741–0911
Fax: (717) 747–7030
inquires@vfis.com
http://www2.vfis.com/

APPENDIX—DOCUMENTATION

This appendix addresses the documentation for the development of the relational firefighter fatality database. An Entity Relational Diagram (ERD) is included to illustrate the relationships used to create the database. Also included is a listing of the field names in each of the database's tables and a brief explanation of the data inside the field. Finally, a data dictionary highlights the valid entries for each data element, with definitions for each of the entries.

Entity Relational Diagram

Figure 29 illustrates the Entity Relational Diagram (ERD) for the Firefighter Fatality database.

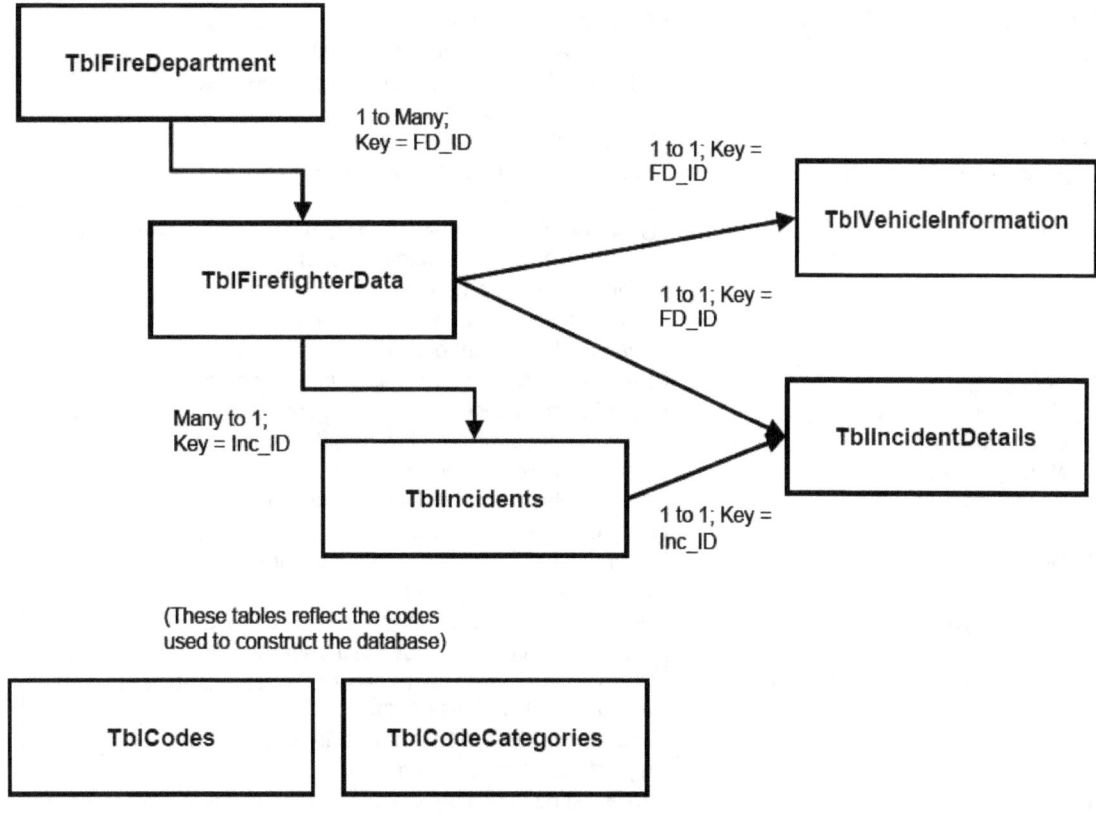

Figure 29. Entity Relational Diagram, Firefighter Fatality Database

Field Names and Description of Data

FIELD NAME	DESCRIPTION
TblCodes	
Code_ID	Structural number used to create the database.
Code_Category	Identifies the data element that the code belongs to
Code_Value	Numeric identifier of data elements for ease of data entry.
Code_Description	Verbal description of code value
Code_Example	Examples to facilitate coding as used in NFIRS coding guide
TblCode_Categories	
Cat_ID	Structural number used to create the database
Cat_Code	Identifies the data element that the code belongs to
Cat_Description	Verbal description of the code values that appear on the Firefighter Fatality Data Collection form
TblFirefighterData	
FF_ID	Decedent identifier used to link information in queries
Inc_ID	Incident identifier, used to link information in queries
FD_ID	Fire department identifier, used to link information in queries
Last Name	Last name of decedent
First Name	First name of decedent
Middle Initial	Middle name or initial of decedent
Rank	Decedent's rank or position in the fire service
Other	Rank not listed above
Gender	Decedent's sex
FF_Year_Birth	Year in which decedent was born
Date of Birth	Decedent's birth date
Affiliation	Decedent's occupational classification with fire department or agency (i.e. career, volunteer, etc.)
FF_Affiliation	Affiliation not listed above
FF_Month_Injury	Month in which decedent's fatal injury occurred
FF_Year_Injury	Year in which decedent's fatal injury occurred
Date of Injury	Date on which decedent's fatal injury occurred
FF_Hour_Injury	Hour in which decedent's fatal injury occurred
Time of Injury	Time of decedent's fatal injury
FF_Month_Death	Month in which decedent's death occurred
FF_Year_Death	Year in which decedent's death occurred
Date of Death	Date of decedent's death
FF_Hour_Death	Hour in which decedent's death occurred
Time of Death	Time of decedent's death
Pre-Existing Conditions	Unused field, all data has been copied to PEC checkboxes
Other	Pre-existing condition not listed under the PEC checkboxes
Smoker within 10 years of death?	Checked if the decedent's records indicate that he smoked within 10 years of his death
Age	Age of decedent at time of fatal injury
FF_Age at Death	Age of decedent at death
FF_PEC_1	Pre-existing condition, checked if the decedent's records indicate that he had a prior heart attack
FF_PEC_2	Pre-existing condition, checked if the decedent's records indicate that he had heart bypass surgery
FF_PEC_3	Pre-existing condition, checked if the decedent's records indicate that he had a cardiovascular condition not described elsewhere
FF_PEC_4	Pre-existing condition, checked if the decedent's records indicate that he had arteriosclerosis

FF_PEC_5	Pre-existing condition, checked if the decedent's records indicate that he had hypertension
FF_PEC_6	Pre-existing condition, checked if the decedent's records indicate that he had diabetes
FF_PEC_7	Pre-existing condition, checked if the decedent's records indicate that he was dehydrated
FF_PEC_8	Pre-existing condition, checked if the decedent's records indicate that he was fatigued
FF_PEC_9	Pre-existing condition, checked if the decedent's records indicate that he had blood clots
FF_PEC_10	Pre-existing condition, checked if the decedent's records indicate that he had a virus or infection
FF_PEC_11	Pre-existing condition, checked if the decedent's records indicate that he had asthma
FF_PEC_12	Pre-existing condition, checked if the decedent's records indicate that he had anemia
FF_PEC_13	Pre-existing condition, checked if the decedent's records indicate that he suffered from seizures
FF_PEC_14	Pre-existing condition, checked if the decedent's records indicate that he had an embolism
FF_PEC_15	Pre-existing condition, checked if the decedent's records indicate that he had an aneurysm
FF_PEC_16	Pre-existing condition, checked if the decedent's records indicate that he had a stroke or cerebral vascular accident (CVA)
FF_PEC_17	Pre-existing condition, checked if the decedent's records indicate that he had pulmonary disease
FF_PEC_18	Pre-existing condition, checked if the decedent's records indicate that he was hearing or vision impaired
FF_PEC_19	Pre-existing condition, checked if the decedent's records indicate that he had a condition that is not listed above, the conditions indicated here are entered into the other field for pre-existing conditions
Incomplete	Checked if information on the decedent is substantially incomplete
TblIncidents	
Inc_ID	Incident identifier, used to link the information in queries
Incident #	Unused field
# of FF Fatalities	Number of firefighters killed in a particular incident
# of FF Injuries	Number of firefighters injured in a particular incident
Summary	Summary of incident from the United States Fire Administration or TriData
TblIncidentDetails	
FF_ID	Decedent identifier, used to link information in queries
Inc_ID	Incident identifier, used to link the information in queries
Type of Duty	Type of duty assignment of decedent (i.e. extinguishing fire, suppression support, en route, etc.)
Other.	Type of duty not classified above
Cause of Fire	Cause of fire that decedent was responding to or involved in
Other.	Cause of fire not classified above
Fixed Property Use	Primary use of facility, structure, or outdoors where incident occurred
Other	Fixed property use not classified above
Type of Incident	Type of emergency that decedent was responding to, returning from, or involved in
Other	Type of incident not classified above
FI_InjuryLocation	Indicates whether decedent was inside or outside of structure when fatal injury occurred

Incident in Vacant/Abandoned Structure?	Indicates whether the incident involved a vacant or abandoned structure
High-rise Incident?	Indicates whether the incident involved a high-rise building
FI_TrainingActivity	Indicates the type of training activity the decedent was involved in, if any
If training activity, what kind?	Other training activity not classified above
Vehicle Accident/Collision Rollover?	Indicates if decedent was involved in a vehicle accident
Alcohol factor in death?	Indicates if records report the presence of alcohol in decedent's blood
Drugs factor in death?	Indicates if records report the presence of legal or illegal drugs in decedent's body
PASS Worn?	Indicates if records mention that decedent was wearing a PASS device
PASS Status?	Indicates if records mention that decedent's PASS device was in the on or off position
PASS Activated?	Indicates if records mention that decedent's PASS device was activated
PPE Worn?	Indicates if records mention that decedent was wearing PPE
SCBA Worn?	Indicates if records mention if decedent was wearing SCBA
Air Supply Depleted?	Indicates if records mention that the decedent was found with his air supply depleted
Equipment Defects?	Indicates if records report any pre-existing problems with decedent's equipment
Type of Defective Equipment	Unused field, all data has been copied to FI_TOE checkboxes
Other	Field for equipment defects not listed under FI_TOE checkboxes
Contributing Factors	Unused field, all data has been copied to FI_CNF checkboxes
Other	Field for contributing factors not listed under FI_CNF checkboxes
Documented Carboxyhemoglobin Level	Carboxyhemoglobin level as reported in decedent's records
Nature of Fatal Injury	The primary apparent symptom the decedent experienced as a result of his fatal injury
Injury Cause Category	Major headings from NFIRS 4.1 cause of injury
Cause of Injury	The cause of a decedent's fatal injury based upon NFIRS 4.1
PSOB Case #	The case number assigned to the claim of a decedent's family under the Public Safety Officer's Benefit Program
PSOB Approval Status	Checked if PSOB benefits were approved. PSOB approves benefits for public safety officers who died as a result of a traumatic injury sustained in the line of duty provided the injury is not self-inflicted and the decedent was not voluntarily intoxicated at the time the injury occurred
Autopsy performed	Indicates whether an autopsy was performed on decedent
Autopsy copy in file?	Indicates whether a copy of the decedent's autopsy is on file
Source of data	Unused field, all data has been copied to FI_SRC checkboxes
Other	Field for contributing factors not listed under FI_SRC checkboxes
FI_CNF_1	Contributing factor to decedent's death as reported by NIOSH or the NFPA, checked if the contributing factor was a human communication error
FI_CNF_2	Contributing factor to decedent's death as reported by NIOSH or the NFPA, checked if the contributing factor was due to a failure of communication equipment
FI_CNF_3	Contributing factor to decedent's death as reported by NIOSH or the NFPA, checked if the contributing factor was due to insufficient resources

FI_CNF_4	Contributing factor to decedent's death as reported by NIOSH or the NFPA, checked if the contributing factor was due to the lack of an accountability system
FI_CNF_5	Contributing factor to decedent's death as reported by NIOSH or the NFPA, checked if the contributing factor was due to the lack of proper incident size up
FI_CNF_6	Contributing factor to decedent's death as reported by NIOSH or the NFPA, checked if the contributing factor was due to the lack of standard operating procedures (SOPs)
FI_CNF_7	Contributing factor to decedent's death as reported by NIOSH or the NFPA, checked if the contributing factor was due to the lack of an effective incident command system
FI_CNF_8	Contributing factor to decedent's death as reported by NIOSH or the NFPA, checked if the contributing factor is not listed above, the data is entered into the other field for contributing factors
FI_SRC_1	Source of data used in the database, checked if the source is the PSOB
FI_SRC_2	Source of data used in the database, checked if the source is the USFA
FI_SRC_3	Source of data used in the database, checked if the source is TriData's files
FI_SRC_4	Source of data used in the database, checked if the source is IOCAD
FI_SRC_5	Source of data used in the database, checked if the source is the NFPA
FI_SRC_6	Source of data used in the database, checked if the source is the NVFC
FI_SRC_7	Source of data used in the database, checked if the source is NIOSH
FI_SRC_8	Source of data used in the database, checked if the source is the IAFF
FI_SRC_9	Source of data used in the database, checked if the source is the Fallen Firefighter's Foundation
FI_SRC_10	Source of data used in the database, checked if the source is from the internet
FI_SRC_11	Source of data used in the database, checked if the source is from the media
FI_SRC_12	Source of data used in the database, checked if the source is the decedent's death certificate
FI_SRC_13	Source of data used in the database, checked if the source is the decedent's autopsy
FI_SRC_14	Source of data used in the database, checked if the source is the decedent's hospital records
FI_SRC_15	Source of data used in the database, checked if the source is from the decedent's fire department.
FI_SRC_16	Source of data used in the database, checked if the source is the decedent's toxicology report
FI_SRC_17	Source of data used in the database, checked if the source is an eyewitness statement
FI_SRC_18	Source of data used in the database, checked if the source is not listed above, the data is entered in the other field for source of data
FI_TOE_1	Type of equipment defect, checked if the defective equipment is the SCBA
FI_TOE_2	Type of equipment defect, checked if the defective equipment is the PASS device
FI_TOE_3	Type of equipment defect, checked if the defective equipment is the radio

FI_TOE_4	Type of equipment defect, checked if the defective equipment is the helmet
FI_TOE_5	Type of equipment defect, checked if the defective equipment is the PPE
FI_TOE_6	Type of equipment defect, checked if the defective equipment is not listed above, the data is entered into the other field for type of equipment defect
TblFireDepartment	
FD_ID	Fire department specific record identifier, used to link the information in queries
FD_Name	Name of the decedent's fire department
FD_City	City where the fire department is located
FD_State	State where the fire department is located
FD_Zip	Zip code where the fire department is located
FD_US_Protectorate	Checked if the fire department is located in a territory of the United States
FD_IncidentZip	Zip code if the incident occurred far from the decedent's home fire department
Community_Population Size	Size of the fire department's response area
Latitude	Latitude of the fire department based on the fire department's zip code
Longitude	Longitude of the fire department based on the fire department's zip code
X-Y Coordinates	Unused field
Type of Agency	Describes the type of organization decedent belonged to (i.e. volunteer, prison brigade, etc.)
Other	Type of agency not listed above
TblVehicleInformation	
FF_ID	Decedent identifier used to link information in queries
Decedent's Status in Vehicle	Indicates whether decedent driving the vehicle or a passenger in the vehicle
Location in Vehicle	Indicates location of decedent in vehicle when injury occurred
VI_LocationInVehicle_Other	Location in vehicle not listed above
Type of Vehicle	Describes the apparatus or POV that decedent was driving or a passenger in
VI_TypeofVehicle_Other	Type of vehicle not listed above
Wearing Seatbelt?	Indicates if records mention that decedent wore a seatbelt
Was Ejected?	Indicates if records mention if decedent was ejected from vehicle
Driver Speeding?	Indicates if the decedent or the driver of the vehicle in which the decedent was riding was speeding

Data Dictionary

This section includes the valid entries for each of the data elements collected in the database, with definitions (where applicable).

Incident Information

Multiple FFF? – Indicates whether more than one firefighter was killed during a particular Incident. If yes, the number of fatalities is indicated.

Firefighter Injuries Number – If firefighters were injured (non-fatally) during the course of the same incident as the fatality, the number of injured firefighters is recorded.

Type of Duty at Time of Injury – Refers to the type of duty the firefighter was performing when he or she was fatally injured.

1	Extinguishing fire/neutralizing incident	Includes handling charged hose lines, operating master stream devices, removing power lines, and other duties directly related to mitigating an emergency incident.
2	Suppression Support	Includes forcible entry, ventilation, water supply, and Incident Command as well as other activities in support of fire suppression activities.
3	Traffic Control	
4	Other duties on Emergency Scene	Miscellaneous duties on the scene of an emergency, not addressed in other categories (e.g., safety officer).
5	EMS	Indicates the firefighter was actually involved in providing patient care at the time of his or her fatal injury.
6	Rescue	Indicates the firefighter was actively engaged in a rescue operation (e.g., extrication) and the time of his or her fatal injury.
7	Inspection	
8	Training	
9	Maintenance	
10	Communications	Includes communications duties at the firehouse or communications center, not on the scene of an emergency incident.
11	Administrative	
12	Prevention	
13	En Route/Responding	
14	Returning/Non-Emergency	
15	In-station duties	Includes normal daily activities, including eating, sleeping, etc.
16	Physical Fitness	
17	Medical Physical	
18	Competition/Muster	
19	Other	Any activity not included in the above categories.

If a fire, Primary Cause – These cause categories were based on the Priority Cause Grouping Code used in NFIRS, but were assigned based on incident reports submitted by fire departments or as reported in the media.

1   Incendiary/Suspicious (Arson)
2   Children Playing
3   Smoking
4   Heating
5   Cooking
6   Electrical Distribution
7   Appliances
8   Open Flame
9   Other Heat
10 Other Equipment
11 Natural
12 Exposure
13 Other

Fixed Property Use of Incident – Location of emergency incident. Incidents where firefighters were involved in motor vehicle collisions are coded based on the incident to which the firefighter was responding.

1 Public Assembly
2 Educational
3 Institutional
4 Residential
5 Store/Office
6 Industrial
7 Manufacturing
8 Storage
9 Street/Highway
10 Water
11 Outdoors
12 In flight/Airport
13 Railway
14 Fire Station
15 Training Facility/Area
16 Other

Type of Incident – The type of emergency incident the firefighter was engaged in neutralizing, responding to, or returning from.

1 Structural Fire/Explosion
2 Wildland/Brush/Grass Fire
3 Nonstructural Fire/Explosion
4 Fire Alarm
5 Motor Vehicle Crash
6 EMS
7 Water Rescue
8 Confined Space/Trench Rescue
9 False Call
10 Training
11 Other

Decedent's Location at time of Injury – Where applicable, whether the firefighter was inside or outside a structure (of any type, residential, commercial, etc.) at the time of his or her fatal injury.

1 Inside Structure
2 Outside Structure
3 Unknown/Not Applicable

Incident in a Vacant or Abandoned Structure? – Whether incident occurred in a vacant or abandoned structure.

1 Yes
2 No
3 Unknown/Not Applicable

High-rise Incident? – Whether incident occurred in a high-rise structure (over three stories).

1 Yes
2 No
3 Unknown/Not Applicable

If a training activity, what kind? – If the firefighter was engaged in a training activity, indicates the general type of training.

1 Live Fire
2 Physical Fitness
3 SCBA

4	Apparatus Drill
5	Driver Training
6	Equipment Drill
7	Underwater/Dive
8	Water Rescue
9	En Route/Returning from Drill
10	Competition
11	Disaster Drill
12	Class/Seminar/Meeting
13	Other

Vehicle Accident/Rollover? – Indicates whether the firefighter was involved in motor vehicle accident or rollover.

1	Yes
2	No
3	Unknown

If YES, Decedent's status in vehicle – If the firefighter was involved in a motor vehicle accident/ rollover, indicates his or her status in the vehicle.

1	Driver
2	Passenger
3	Unknown

Location in Vehicle – Indicates the firefighter's location in the vehicle at the time of injury.

1	Drivers' Seat
2	Front Passenger
3	Rear Passenger
4	Jump Seat
5	Back Step
6	Payload Area
7	Outside Vehicle
8	Other

Type of Vehicle – Indicates what type of vehicle the firefighter was driving or riding in at the time of injury.

1	Engine/Pumper
2	Tanker
3	Personally Owned Vehicle (POV)
4	Ladder Truck/Aerial Tower
5	Heavy Rescue
6	Ambulance
7	Brush Truck
8	Pickup Truck/4X4
9	Fire Dept Car
10	F.D. Display/Competition Vehicle
11	Boat
12	Helicopter
13	Airplane
14	Other

Decedent wearing seatbelt? – Indicates whether the firefighter was reportedly wearing a seat-belt at the time of injury.

1	Yes
2	No
3	Unknown

Decedent Ejected? – Indicates whether the firefighter was ejected from the vehicle.

1   Yes
2   No
3   Unknown

Driver Exceeding Speed Limit? – Indicates whether the vehicle driver was known to have exceeded the speed limit prior to the fatal collision or rollover.

1   Yes
2   No
3   Unknown

Alcohol a factor in death? (Decedent's BAC/BAL) – Indicates whether the firefighter's blood alcohol level was reported to have been a factor in his or her death.

1   Yes
2   No
3   Unknown

Drugs (prescription/illegal) – Indicates whether the firefighter's ingestion of drugs, prescription or illegal, were reported to have been a factor in his or her death.

1   Yes
2   No
3   Unknown

PASS Worn? – Indicates whether the firefighter was reportedly wearing a PASS Device at the time of injury.

1   Yes
2   No
3   Unknown
4   Not Applicable

PASS Status? – If worn, indicates the status of the firefighter's PASS device, when the firefighter was found.

1   On
2   Off
3   Unknown
4   Not Applicable

PASS Activated? – If worn, indicates whether the firefighter's PASS device reportedly activated.

1   Yes
2   No
3   Unknown
4   Not Applicable

PPE Worn? – Indicates whether the firefighter was wearing PPE appropriate to the incident at the time of injury.

1   Yes
2   No
3   Unknown
4   Not Applicable

SCBA Worn? – Indicates whether the firefighter was wearing and SCBA at the time of injury.

1   Yes
2   No
3   Unknown
4   Not Applicable

Air Supply Depleted? – If the firefighter was wearing an SCBA, indicates whether its air supply was found to have been depleted.

1   Yes
2   No
3   Unknown
4   Not Applicable

Equipment Defects or failures noted? – Indicates whether equipment defects or failures were noted in media reports, fire department incident reports, or other incident investigations.

1   Yes
2   No
3   Unknown
4   Not Applicable

If YES, which piece of equipment? – Indicates which piece of equipment reportedly failed or was defective.

1   SCBA
2   PASS
3   Radio
4   Helmet
5   PPE
6   Other

Contributing Factors (NIOSH/NFPA) – Indicates contributing factors identified by the NFPA, NIOSH or other investigating agencies.

1   Human Communication Error
2   Communication Equipment Failure
3   Insufficient Resources
4   Lack of Accountability System
5   Lack of Proper Incident Size
6   Lack of SOPs
7   Lack of Effective Incident Command System
8   Other

Firefighter Information

Name

Date of Birth

Gender

1   Male
2   Female
3   Unknown

Date of Injury

Time of Injury

Age at Injury

Date of Death

Time of Death

Age at Death

Rank – Firefighter's rank at time of injury.

1   Chief
2   Assistant/Deputy Chief

3   Battalion/District Chief
4   Captain
5   Lieutenant
6   Sergeant
7   Engineer
8   Firefighter
9   Paramedic
10   EMT
11   Pilot
12   Forestry Worker
13   Forestry Technician
14   Fire Police Officer
15   Auxiliary Member
16   Fire Marshal
17   Inspector
18   Junior Firefighter
19   Recruit/Probationary
20   Other

Affiliation – Firefighter's affiliation at the time of injury.

1   Paid (Full-time)
2   Paid (Part-time)
3   Volunteer
4   Contract
5   Inmate/Prisoner
6   Military
7   Wildland (Full time)
8   Wildland (Fart time)
9   Other

Pre-existing Conditions (all that apply) – Conditions noted on autopsy reports, by family members, by other firefighters, or on other documents associated with the fatality.

1   Prior Heart Attack
2   Heart Bypass
3   Cardiovascular Other
4   Arteriosclerosis
5   Hypertension
6   Diabetes
7   Dehydration
8   Fatigue
9   Blood Clots
10   Virus/Infection
11   Asthma
12   Anemia
13   Seizures
14   Embolism
15   Aneurysm
16   Stroke/CVA
17   Pulmonary Disease
18   Hearing/Vision Impaired
19   Other

Decedent a smoker within 10 years of death? – Based on autopsy reports, reports from family members and other firefighters, or on other documents associated with the fatality.

1   Yes
2   No
3   Unknown

Fire Department Information

Fire Department Name

Community Size (Population)

Location (City/State/ZIP

US Protectorate?

    1   Yes
    2   No

Latitude

Longitude

X-Y Coordinates

Type of Agency – Agency with which the firefighter was affiliated at the time of injury.

    1   Career
    2   Volunteer
    3   Combination
    4   Federal
    5   Private Contractor
    6   Prison Brigade
    7   Other

Injury Information

Nature of Fatal Injury – Generally taken from autopsy records or incident reports; reflects the primary nature of the firefighter's fatal injury.

    1   Amputation
    2   Asphyxiation
    3   Aneurysm
    4   Bleeding
    5   Burns
    6   Burns/Asphyxiation
    7   Cancer
    8   Cardiac Arrest/Heart Attack
    9   Chemical Exposure
    10   Dehydration
    11   Diabetic Coma
    12   Drowning
    13   Electric Shock
    14   Gunshot
    15   Infectious Disease
    16   Internal Trauma
    17   Overdose
    18   Other Medical
    19   Respiratory Arrest
    20   Stroke/CVA
    21   Other cardiovascular
    22   Puncture Wounds/Stabbing
    23   Undetermined

Documented carboxyhemoglobin level – If available from autopsy records, reflects the level of carboxyhemoglobin in the firefighter's blood.

Cause of Firefighter Injury – Reflects the proximate cause of the firefighter's fatal injury.

1. Fell/Slipped	
101	In hole, outside structure.
102	In hole burned in floor
102	In hole burned in roof
103	In unguarded opening in floor.
104	In unguarded opening in roof.
105	Over object.
106	On wet surface.
107	On icy surface.
108	On flat surface.
109	On steps/stairs.
110	From ladder.
111	From structure.
112	From emergency apparatus (safety bar/belt not fastened).
113	From emergency apparatus (safety bar/belt failed).
114	In or into emergency apparatus.
115	Off station sliding pole.
116	Over/off curb.
199	Fell/Slipped not classified above.
100	Fell/Slipped; insufficient information to classify further.

2. Caught/Trapped — In, By, Between	
201	Collapsing roof.
202	Collapsing wall.
203	Collapsing floor.
204	Collapsing ceiling.
205	Fire progress.
206	Back draft.
207	Flashover
208	Explosion.
209	Falling object(s).
210	Between objects.
211	Lost inside building.
212	Fire department apparatus.
213	Earth cave-in.
214	Underwater object(s) or obstacle(s).
299	Caught/Trapped not classified above.
200	Caught/Trapped; insufficient information to classify further.

3. Struck By	
Use when injury-producing object moved toward injured person.	
301	Collapsing roof.
302	Collapsing wall.
303	Collapsing ceiling.
304	Pieces of wall.
305	Ceiling being pulled by self.
306	Ceiling being pulled by others.
307	Dirt particles.
308	Flying glass.
309	Glass broken by self.
310	Glass broken by others.
311	Water stream, hand line.
312	Water stream, master stream.

313	Portable extinguisher stream.
314	Ladder.
315	Hand tools/equipment.
316	Hose.
317	Coupling.
318	Fire department apparatus.
319	Non-fire department vehicle.
320	Falling object(s).
321	Thrown objects, non-malicious (see 803 for Malicious).
322	Apparatus and compartment doors.
323	Other doors.
324	Fire station equipment.
325	Other personnel (not intentional).
399	Struck By to classified above.
300	Struck By; insufficient information to classify further.

4. Contact with/Exposure to

401	Heat.
402	Embers.
403	Hot metal.
404	Hot tar, etc.
405	Fire.
406	Splinters.
407	Nails.
408	Glass.
409	Water.
410	Steam.
411	Smoke/toxic fire products.
412	Unusual fumes, gases.
413	Chemicals.
414	Radioactive material.
415	Electricity.
416	Utility flames, flairs, torches, etc.
417	Underwater objects.
418	Insects.
419	Poisonous plants.
420	Contagious disease.
421	Extreme weather.
499	Contact with/Exposure to not classified above.
400	Contact with/Exposure to; insufficient information to classify further.

5. Overexertion/Strain

Use if related to the activity at the time injury is received.

501	While lifting hose.
502	While lifting ladder.
503	While lifting hand tools, saws, etc.
504	While lifting victim (during rescue from fire operation).
505	While lifting property/contents.
506	While lifting during rescue operation.
507	While lifting other, not classified above.
508	While carrying hose.
509	While carrying ladder.
510	While carrying hand tools, saws. etc.
511	While carrying victim (during rescue from fire operation).
512	While carrying property/contents.
513	While carrying during rescue operation.
514	While carrying, not classified above.
515	While pulling hose.

516	While pulling ladder.
517	While pulling hand tools, saws, etc.
518	While pulling victim (during rescue from fire operation).
519	While pulling property/contents.
520	While pulling during rescue operation.
521	While pulling other, not classified above.
522	While in rescue operation.
523	While climbing ladders.
524	While climbing stairs.
525	While climbing cliff or wall.
526	While climbing not classified above.
599	Overexertion/Strain not classified above.
500	Overexertion/Strain; insufficient information to classify further.
6. Exiting or Escaping — Jumped	
601	From ladder.
602	From wall, ledge, or window.
603	From roof.
604	From other part of structure.
605	From fire department apparatus.
699	Exiting or Escaping — Jumped not classified above.
600	Exiting or Escaping — Jumped; insufficient information to classify further.
7. Fire Department Apparatus Accident	
701	Collision with other vehicle.
702	Collision with pedestrian
703	Collision with stationary object(s).
704	Collision not classified above.
705	Collision; insufficient information available to classify further.
706	Left road (no collision).
707	Overturned (no collision).
799	Fire Department Apparatus Accident not classified above.
700	Fire Department Apparatus Accident; insufficient information available to classify further.
8. Assaulted	
801	Struck by individual(s) (deliberate act).
802	Struck by individuals (crowd action pushing or shoving).
803	Struck by thrown object(s).
804	Cut/stabbed.
805	Gunshot.
806	Bitten.
899	Assault not classified above.
800	Assault; insufficient information available to classify further.
9. Other Cause	
999	Cause of Fire Fighter Injury not classified above.
000	Cause of Fire Fighter Injury undetermined or not reported

Miscellaneous

PSOB Case #

PSOB Approval Status

1   Approved
2   Denied
3   Unknown, Other

Autopsy performed?

1   Yes
2   No
3   Unknown

Copy of Autopsy report in file?

1   Yes
2   No
3   Unknown

Brief summary of incident – Taken from media reports, incident reports, investigations, and other records.

Source(s) of Data

1   PSOB
2   USFA
3   TriData
4   IOCAD
5   NFPA
6   NVFC
7   NIOSH
8   IAFF
9   Fallen Firefighters' Foundation
10 Internet
11 Media
12 Death Certificate
13 Autopsy
14 Hospital Records
15 Fire Dept. Report
16 Toxicology Reports
17 Eyewitness Statement
18 Other

Rules for Coding Data

The following are general rules for common types of incidents. In incidents with unusual circumstances or where the events did not exactly conform to the coding categories, it is the coder's discretion how to interpret the data.

Required Fields

Name

Date of Injury

Nature of Injury

Residential Structure Fire

FPU = 4

TOI = 1

Motor Vehicle Collision (involving fire apparatus)

FPU = For incident to which the FF was responding

TOD = 13

Vehicle Accident/Rollover = YES (along with accompanying information, e.g., status in vehicle)

COI = 7xx

Motor Vehicle Collision (involving POV)

FPU = For incident to which the FF was responding

TOD = 13

Vehicle Accident/Rollover = YES (along with accompanying information, e.g., status in vehicle)

COI = 999

Motor Vehicle Collision as Incident

FPU = 9 (generally)

TOI = 5

EMS Call

TOI = 6

Other Rules

No entries should have "nulls." Entries should be coded as "unknown" or not applicable/available.

Ensure that the agency and affiliation for each firefighter are consistent (e.g., generally, volunteer firefighters should be affiliated with either volunteer or combination agencies).

States and ZIP codes must match (to facilitate geocoding the data).

Where possible, record both the state where the firefighter was affiliated and the State where he or she died (this is especially important for wildland firefighters).

For heart attacks, COI should be coded as 5xx if the firefighter was actively engaged in some kind of physical activity at the time of his or her death. For those firefighters that were engaged in normal station duties or other administrative tasks, COI should be coded as 999.

REFERENCES

1 28 CFR § 32.2(g)(2001).

2 "5-Year Operational Objectives," U.S. Fire Administration, October 2000.

3 Incident data courtesy of the National Fire Protection Association.

4 Karter, Michael, *U.S. Fire Department Profile Through 1999*, National Fire Protection Association, 2000.

5 *Comparison of Volunteer and Career Municipal Firefighters*, U.S. Fire Administration, August 1992.

6 *Training Fatalities 1978–1987*, United States Fire Administration, August 1988.

7 *Fatalities Involving Motor Vehicle Accidents 1984–1993*, National Fire Protection Association.

8 *Fire in the United States 1989–1998, 12th Edition*, U.S. Fire Administration, Federal Emergency Management Agency, August 2001.

9 *Multiple Firefighter Fatalities, 1982-1991*. U.S. Fire Administration, August 1992.

10 National Institute for Occupational Safety and Health, Reports F200–23, F2000–16, and 99F–48.

11 NFPA 1583, *Standard on Health-Related Fitness Programs for Fire Fighters*, National Fire Protection Association, August 2000.

12 Womack, Wade, *Cardiovascular Risk Markers in Firefighters: A Longitudinal Study,* Applied Exercise Laboratory, Texas A&M University, May 2001.

13 Williford, Henry N. and Michele Scharff-Olson, "Fitness and Body Fat: An Issue of Performance." *Fire Engineering*, August 1998.

14 "Oklahoma Department Fights Fat, Fatigue," *Firehouse.Com News*, May 24, 2001.

15 Fire Service Joint Labor Management Wellness-Fitness Initiative, International Association of Fire Fighters.

16 *Heart Attack Treatment, Recovery and Prevention,* American Heart Association, 1998. Also see: *Sudden Cardiac Death,* American Heart Association, 2000.

17 Office on Smoking and Health, National Center for Chronic Disease Prevention and Health Promotion, Centers for Disease Control.

18 *Health, United States: 2000*, Center for Disease Control and Prevention, 2001.

19 NFPA 1403, *Standard on Live Fire Training Evolutions in Structures,* National Fire Protection Association.

20 NFPA 1521, *Standard for Fire Department Safety Officer,* National Fire Protection Association.

21 NFPA 1982, *Standard on Personal Alert Safety Systems,* National Fire Protection Association.

22 Burgess, J. L., et al., "Adverse Respiratory Effects Following Overhaul in Firefighters." *Occupational Environmental Medicine*. May 2001; 43 (5): pp. 467-473.

23 NFPA 1002, *Fire Department Vehicle Driver/Operator Professional Qualifications,* National Fire Protection Association.

24 Biggers, W. A., Jr., Zachariah, B. S. and Pepe, P. E., "Emergency Medical Vehicle Collisions in an Urban System." *Prehospital Disaster Medicine 11* (3): 195-201, 1996.

25 Fahy, Rita and Paul R. Leblank, "U.S. Firefighter Fatalities: 2000," *NFPA Journal,* July/August 2001, Vol. 95 (4): p. 67–79.

www.ingramcontent.com/pod-product-compliance
Lightning Source LLC
Chambersburg PA
CBHW081214170526
45165CB00009B/2816